THE LORE OF
STEAM

C. HAMILTON ELLIS

Hamlyn Paperbacks

THE LORE OF STEAM

ISBN 0 600 20783 8

First published in Great Britain 1984
by Hamlyn Paperbacks
World copyright © AB Nordbok, Gothenburg, Sweden,
1975 and 1983

Hamlyn Paperbacks are published by
Arrow Books Limited
17-21 Conway Street, London W1P 6JD, England
A division of the Hutchinson Publishing Group
London Melbourne Sydney Auckland Johannesburg
and agencies throughout the world

The Lore of Steam was, designed and produced by
AB Nordbok, 1983

Typesetting by Text Processing, Clonmel, Ireland
Printed in Spain 1983 by Novograph, S.A., Madrid

Also available from Hamlyn Paperbacks:
THE LORE OF SAIL

Contents

Preface

With a life-long experience of working on trains behind me I feel qualified to express my admiration for the way in which the author and the publishers have dealt with The Lore of Steam.

"Lore", meaning knowledge gained by study, symbolizes the author's thoroughness in dealing with his subject and it is quite obvious that it must have grown up with him from boyhood. The vast amount of knowledge and information he has imparted, relating to travel in so many parts of the world, is truly remarkable. This I can substantiate from my own personal experiences. The illustrations are excellent and portray the atmosphere that covers so many climates.

Hamilton Ellis proposed that the preface should be written by one of the family that started the whole business of powered railroad traction, namely the Trevithicks. It was agreed, and he wrote to me: "The Americans will not mind that you are not descended from Oliver Evans."

As the reader will discover, all this began with strong steam, which was pioneered by Richard Trevithick before 1800. This high-pressure steam enabled the steam engine to move about under its own power, and it thus became a prime mover. Steam engines have served the world of travel on both land and sea in very good stead for a matter of 150 years and although the steam locomotive is practically extinct, it is not so with strong steam as it is now being used in ever-increasing quantities for the generation of electricity, which in turn propels so many of the world's electric trains.

I believe that those interested in trains – past, present and future – will find a mine of information in this book.

RICHARD TREVITHICK
Dorset, England

Chapter 1
Primeval Forms

What is a train?

Well, the word, or its equivalent, means several things in many languages. It is almost the same in English, French, Dutch, Spanish and Italian, and can mean anything from the tail of a ceremonial garment held up by pages behind an important personage to a complete procession of people and animals. German and Scandinavian languages produce Zug, Tåg and Tog, which are practically the same as the English verb tug, meaning to pull hard. The thing has a fascinating etymology.

In the general usage of our time, the word means some sort of multiple vehicle, propelled by mechanical means, and running on a strictly confined track. So much for the word.

This sort of train has done more to change the world of mankind than any other form of transport since the canoe evolved into the ship. It did for the Continental land-masses what the ship, centuries before, had done for the oceans. After it came the motor, and then the aircraft, which could overcome both sea and land distances in very fast traffic. But between them, the ship and then the train opened up the world.

At a remote time, someone invented the wheel. At a later time, but still remote, someone else invented the guided vehicle. Both events probably took place somewhere in Western Asia, in Mesopotamia, the country of the great rivers.

The guided vehicle?

That was the wheeled vehicle whose course was determined by the ruts it ran in. Ancient vehicles had gouged out those ruts in the streets of Ur, of Babylon, and in the ancient cities of Assyria. Men soon saw that the ruts, once they had formed, kept the vehicles to a fixed path, so that they damaged neither themselves nor the corners of the buildings round which they passed in the close-built riverside cities along the Tigris and the Euphrates. Then paving succeeded the mud roads and in that paving, ruts were made deliberately by measure, so that the carts still should follow a disciplined course. When the four-wheeled cart was a new invention, there were no swinging front axles!

It was not the beginning of railways, but it was the beginning of the Railway Idea. The Prophet Isaiah, who was a very discerning man with things as well as people, knew more than a thing or two about the Assyrians and their ways, when back in the eighth century B.C. he wrote about the crooked being made straight and the rough places plain. He had seen their marvellous stoneways, and he looked into the future, when every valley should be exalted and every mountain and hill laid low.

The Aelopile, a primitive form of steam reaction turbine ascribed to Hero of Alexandria. He probably considered it to be nothing more than an amusing plaything.

Those stoneways, as suggested, were not railways, but they were tramways of a sort. They were invaluable for the movement of huge blocks of stone from quarry to city, or of religious monuments from quarry or city to some holy place, and for these things they were used by the Greeks, who took the idea with them when they colonized Sicily, whence the Romans copied them. One can still see stone rut-ways in the quarries of Syracuse, which served Dionysius Tyrannus both for building material and as a place to keep political prisoners and captured soldiers out of mischief. One can see them in the relatively modern paved streets of Pompeii, volcanically overwhelmed and thus by the irony of history preserved, in 79 A.D.

The Greeks seem to have got as far as making turnouts and passing loops, the remote precursors of the points or switches in a modern railway. The Greek word for them was *ektropoi*. But there was no mechanical agent beyond that of the lever and the pulley. Motive power was furnished by men (probably poor devils of slaves or convicts), by horses, asses, mules, or possibly camels in some places.

Yet the thing — the Railway Thing — almost might have happened even then. Hero, an Alexandrian Greek, devised a primitive reaction turbine — a steam-filled whirling ball with opposed escape-jets — about this time. It was simply a scientific toy, though something like it is said to have been used for turning spits in medieval kitchens, notably in the great monasteries where learning was kept alive in the Dark Ages.

Mining railway illustration of about 1519 from Der Ursprung gemeynner.

The Romans, like the Minoans of Crete before them, were great plumbers and domestic engineers, and one of their appliances was an internally-fired boiler, heated by burning charcoal in an extraordinarily advanced water-tube firebox. But its service was only that of the bathroom in the houses of the rich. Nobody bothered about motive power for industrial machines and transport. There were plenty of slaves and prisoners-of-war around, as well as animals, for such power as people thought they needed.

Classic Greece declined into a Roman satellite state, ultimately to fall to the Turks. What we now call the Middle East, with the past glories of Sargon, King of Kings, buried in the sand, furnished outposts for the Roman Empire. Nor did Imperial Rome complete her civilization of Europe and Western Asia, though, to be sure, her stoneways as well as her ordinary roads stretched far and wide, to be regarded and then forgotten by invading Celt, Goth, Frank, Saxon, Hun and Turk. For Imperial Rome died as Assyria and Persia and Macedon had died, and the Dark Ages set in for many countries over many centuries.

They were ages so dark that only in the nineteenth century did people of Western Europe unearth the remains not only of well-planned cities and garrison towns, but of splendid country houses, centrally heated and admirably drained; of beautifully paved roads deep down below muddy pack-horse lanes, and, rather significantly, those rutted stoneways whereon wagons could pass with heavy loads yet without need of steering. One of these turned up in the British Isles, with quite curious significance, on the site of Abbeydore railway station on the border of England and Wales.

That, then, was the beginning of the guided vehicle, but not of the railroad proper. Whence came that? The most facile explanation is that on rutted roads in wet months men laid split tree-trunks in the bottoms of the muddy ruts to sustain the wheels, and that these constituted the first real rails. No doubt this was done in many places where there was little stone and plenty of mud. But the essence of the railway was the use of flanges, either in the track or on the wheels themselves. The stoneways had provided flanges in the form of their track. But what about the flanged wheel on the plain rail?

In its most primitive form, this was produced by an arrangement of great grooved bobbins to act as wheels for the wagons, while smoothly-trimmed tree-trunks—spruce or larch for example—laid on and secured to much shorter sections at right-angles, formed the track. Thus there were rails on sleepers or cross-ties, and there were even primitive switches. Just who first made such a track no man dare say, but early in the sixteenth century it was certainly being used by the miners in the gold-diggings of Transylvania, and specimens of both track and vehicle have almost miraculously survived the centuries.

We find illustrations of early wooden wagonways in several sixteenth-century treatises, of which perhaps the best-known is the *De Re Metallica* of Georgius Agricola. It was published in 1556. Six years before, a mining railway in Alsace was illustrated by Sebastian Münster in his *Cosmographiae Universalis*. Mining railways in Eastern Europe, and also in the Tyrol, probably had been in use for quite a time before these publications.

Newcomen's low-pressure steam engine for pumping water out of mines. It was built in the early 18th century and was the first commercial use of steam.

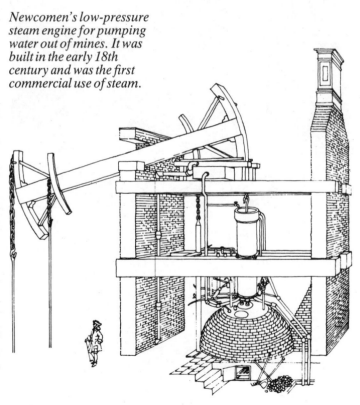

Virgula divina

Glück rhe

Haspeler

Instrumentum Tractorium

Bünnan

Zerseczer

Scübever

Häuwer

(Below) *Wagon on flanged rollers in use in South-east Europe in the early 16th century.*

(Above) *This print, from Sebastian Münster's* Cosmographiae Universalis (1550), *shows a mining truck with iron wheels.*

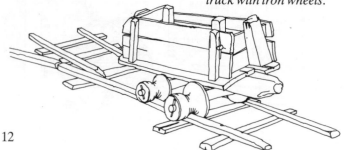

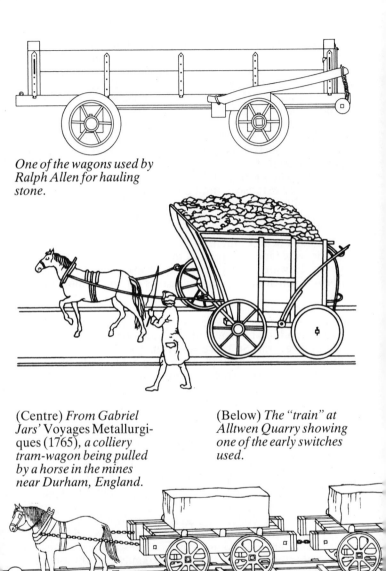

One of the wagons used by Ralph Allen for hauling stone.

(Centre) *From Gabriel Jars'* Voyages Metallurgiques (1765), *a colliery tram-wagon being pulled by a horse in the mines near Durham, England.*

(Below) *The "train" at Alltwen Quarry showing one of the early switches used.*

One can only write "probably", noting that as far as we know the idea of the confined track occurred to some unrecorded Mesopotamian, and the use of flanged wheel on plain rail to some forgotten German, not necessarily in the countries we now call Iraq and Germany. The wagons were called in Old German *Hunte* (dogs).

In the eighteenth century, there were two rival systems; that of the flanged wheel on the plain rail, the present form, and that of the plain wheel on the flanged rail, the latter being formed of L-shaped iron plates extending from stone to stone; a rough but workable road so long as loads were not too heavy. In that century, very extensive mining railways were built in Europe, most notably in South Wales and North-eastern England, where collieries were booming. In Scotland, too, they had arrived sufficiently soon for a battle to be fought over one (at Prestonpans near Edinburgh in 1745) during the last British dynastic war. One may read of these and many other things in Charles Edward Lee's scholarly book *The Evolution of Railways*, which quite properly leaves off where most railway books begin.

What, one may ask, has all this to do with trains? A horse pulling a wagon over rails did not make a train. But while, on a common road, that horse could pull but one wagon, on rails he could pull several. So there was your train!

Still, it could not be our idea of a train until it had a propulsive engine of some sort. In the sixteenth and seventeenth centuries, there were no engines apart from very simple and quite academic models. What in later years was to be called Civil Engineering (that was, not military as in the building of fortifications) was a much older science than that of mechanical engineering. The Roman aqueducts built under the emperor Claudius were quite "modern".

In eighteenth-century Europe, particularly in North-eastern England, there were tremendous earthworks to support the archaic mining railways. The mines were often in the hills, and the railways went down to the nearest waterway where ships could come in to collect the coal. Going uphill, the horse or horses pulled the wagons. Downhill they rode in a "dandy-cart" attached at one end. So used to the drill were the horses that they would trot round and mount the dandy-cart without being led. For the rest, movement was controlled by brake levers and blocks on the wheels.

There still stands in County Durham, England, what is probably the first railway viaduct in the world. It is the magnificent Tanfield Arch, built in 1727. The mines it served

14

were probably worked out by the end of the century, but like the Roman aqueducts it still stands, and under the British Government today it is scheduled as an ancient monument.

Mechanical engineering also had much to do with military matters. It is not absurd to suggest that the first spring was the bow, and that the most primitive internal-combustion engine was the cannon. But just as mining produced the first railed tracks, so did mining produce the first steam engines. Their work was to pump out the water that was always flooding the workings. In the seventeenth century, the Marquis of Worcester in England and Denis Papin and Thomas Savery in both France and England (Papin lived for some years in Kassel) made various applications of steam power through displacement.

Early in the eighteenth century, in England, Thomas Newcomen made the first commercially practical reciprocating steam engines, for pumping water from mines. The real power was that of the atmosphere, for steam was admitted simply to create a vacuum in the bottom of the cylinder by condensation. James Watt (1736-1819), a Scot, improved on Newcomen's system by providing a separate condenser, greatly speeding the action of the engine, and also produced the first practical rotary motion by crank and flywheel (initially with the "sun-and-planet" gear, as the crank was covered by another man's patent). Watt's engine could drive machines as well as work a pump. Pressure was still very low.

Before that, Papin had produced a high-pressure boiler, though in a simple form, such as we could call a pressure-cooker ("Papin's Digester").

Brunton's locomotive, 1813. This strange-looking object was supposed to push itself along by its "hind legs".

At the end of the eighteenth century came Richard Trevithick, a Celt from Cornwall, where he was an official and engineer in one of the tin mines. He it was who applied high-pressure steam to the driving of a double-acting reciprocating engine. The slide valve for steam admission and exhaust had been invented by one of James Watt's young men, William Murdoch. The Trevithick engine was small, compact, and immensely powerful for its size. It was the answer to the industrialists' prayer. The huge low-pressure engines of Watt had needed a large building to house them, but one could put a Trevithick engine almost anywhere.

Watt was much annoyed and, playing on the idea that the use of high-pressure steam was very dangerous, publicly remarked that his rival ought to be hanged. But at last there was a steam engine sufficiently compact to make the locomotive a practical possibility.

Trevithick's Black Billy *was in use in the collieries of Northumberland, England, for many years.*

Attempts there had been already. In the late seventeen-sixties Nicolas Cugnot, a French artillery officer, had produced a steam-wagon for gun traction on roads, propelled by a kettle-shaped boiler supplying steam to a pair of single-acting cylinders which drove a single front wheel through ratchet and cog. It did move itself, but was unmanageable as well as being able to steam only for a short while.

Murdoch, too, had experimented with a very small model locomotive (again not for rails) fired by a spirit lamp. This ran quite well; it was a most engaging steam toy. Murdoch's chief, James Watt, was not amused.

Trevithick, however, approached steam locomotion very seriously, first with a model, and then with full-size road locomotives, basing them on his existing stationary engine which had been put already to commercial use. With his partner Vivian in Cornwall, he built a steam road carriage which they drove along the roads from Redruth to Plymouth in 1802, there shipping it to London where it was demonstrated to an unenthusiastic public. It was in fact the first practical motor-car, and people were not yet ready for motor-cars. Further, it got damaged, and now became apparent a fatal defect of Trevithick's character.

He was a giant among original inventors and mechanical engineers. He was on the other hand about as poor a businessman as ever hoped to exploit an invention. If anything broke, or otherwise went wrong, he lost interest in it and went after something else, like an artist burning an unsuccessful picture. But though the steam carriage was lost, all was not lost. The world's first railway locomotive was about to be born.

In the winter of 1803-4, Trevithick was in South Wales, and it was there that this engine was built, to be demonstrated on the Penydarren mining railway, or tramroad as it was called, near Merthyr Tydfil. On February 21, 1804, it was publicly steamed. Even in stern, evangelical Wales, it was a time of heavy betting. The owner of Penydarren Ironworks wagered a neighbouring ironmaster that the "travelling engine" would haul ten English tons of iron on the tramroad from Penydarren to Abercynon, a distance of 9¾ miles. The stakes were 500 guineas (£525 sterling in gold). Trevithick's locomotive made the journey in four hours, five minutes, with the stipulated load of pig-iron, on top of which about seventy men had climbed to enjoy such a novel experience, that of being the first people to travel by a mechanically powered railway train. The track was of flanged iron rails on stone blocks, the common form of the period.

Original drawings have not survived, at any rate entire, but from certain very old drawings contemporary with Trevithick, we know enough to make a very close reconstruction of the engine.

It was in all important features a locomotive version of the Trevithick stationary engine, with a single cylinder driving a transverse shaft with a very large flywheel. The boiler was internally fired, with a return flue. Our drawing is sufficiently explanatory of the way in which the power was transmitted from this shaft, through spur wheels, to the four wheels on which the engine was carried. The single cylinder was embedded in the boiler above the furnace and flue, with the piston rod issuing at one end to drive a crosshead on two parallel slidebars, whence the connecting rods went back to cranks on the transverse shaft at the other end. With little doubt, the crosshead and slidebars were a contribution of William Symington, the Scots pioneer of steam navigation (1763-1831).

Not only is the actual performance of Trevithick's locomotive on record; so is her fuel consumption. Two hundredweight of coal sufficed on the famous trial run.

Did Richard Trevithick press home this advantage with a rich patron who had backed his engine with such a princely wager?

Well he might have done, but the wonder was a thing of a few months only. For the first time, but not for the last, it was seen how an iron locomotive smashed to pieces the flanged cast-iron plates which formed the early rails. A sufficiently substantial wooden way, even, would have answered better.

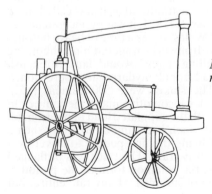

Murdoch's steam carriage, 1784.

18

Yet Richard Trevithick—"Captain Dick" as his Cornish friends called him—had made a contribution to technological history anticipating, in magnitude, the first flights of the Wright Brothers in the United States, ninety-nine years after. Here and there an odd Trevithick locomotive cropped up. There was one, locally known as *Black Billy*, in the Northumbrian coalfields of North-eastern England, which seems to have been an unfortunate venture. It was doubtless seen there by George Stephenson and other torch-bearers of steam locomotion.

In London again, Trevithick showed yet another steam locomotive he had built, called *Catch-me-who-can*, on a circular track in the northern suburbs of that time (1808) near the present Euston Road. Round the circle was a high fence, to guard the thing against becoming a free show. The curious were admitted at a shilling a head, which covered a ride for the more adventurous ones in an adapted carriage drawn by the little locomotive. The latter appears, from an old drawing, to have been smaller than the Penydarren engine, without the big flywheel and complex gears; a sort of mechanical horse on rails. The outfit was indeed a sort of circus, but be it remembered that the visitors were truly the first fare-paying passengers ever to have been hauled on the rail by any sort of locomotive.

There was no high-powered advertising in those days; the newspapers were scornfully indifferent. Before long, there was a derailment. It does not seem to have been serious, with anybody badly hurt, or we should have heard more of it from people who detested the mere existence of any sort of engine.

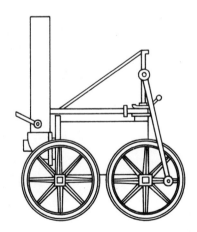

Catch-me-who-can *was exhibited by Trevithick and his partner near Euston Square, London between July and September, 1808 in an attempt to attract financial backing. The engine weighed 8 tons and ran at 12 m.p.h. round a circular track.*

But it was enough for the ingenious inventor once more to lose interest. Very possibly he was too hard-up to repair the damage. He was often so.

Thus the true father of the powered railway train passes out of its history. Yet not only is that to be claimed for him. He was the father of all mechanical land transport, for his steam road carriage of 1802, unlike Cugnot's courageous but quite unmanageable machine, was the first of its kind to carry people on a real journey.

The Stephensons working on the Northumbrian *shortly before the opening of the Liverpool and Manchester Railway. Robert is firing while his father is oiling. The top hat was regarded as appropriate dress for such work. After a drawing by Alexander Nasmyth.*

Chapter 2
The Motive Revolution

So far we have seen the coming of the three essentials; firstly, the principle of the guided vehicle going back to remote times; then, the industrial use of vehicles on raised rails, which certainly was known at the end of the Middle Ages; and thirdly, the dawn of the powered vehicle, which preceded by one year the defeat of Napoleonic sea power off Cape Trafalgar. Indeed, the Napoleonic phase of Western politics just about coincides with the first rise of industrial mechanism and the application of the engine to transport.

One should remark that the steam ship and the locomotive appeared *almost* simultaneously, but owing to the confined dimensions imposed by conditions of land movement, the steam ship made the more rapid start, just as the stationary steam engine had been in advance of both. Regarding a locomotive of the late eighteen-twenties, and then a mill engine of the same time, we find the latter almost "modern" by comparison. There are many "almosts" in this, as in many other phases of mechanical history.

In 1804, the year in which Trevithick had produced the world's first railway locomotive of any sort, Oliver Evans in the United States made America's first locomotive vehicle which, though it had no connection with railroads, strangely anticipated the amphibious motor vehicles which were to serve military purposes nearly a century-and-a-half later. Evans was a gifted and ingenious character who, significantly, plied the trades of both boatbuilder and blacksmith in Philadelphia, Pennsylvania. There he built a remarkable punt-like craft which he named *Oructor Amphibolis*. Not for him the quite skittish nomenclature of his contemporary Trevithick! In America, however, the machine was more generally remembered as "Evans' Scow". In the big punt he mounted a small steam beam-engine, geared to a little paddle-wheel astern. But to get it down to the Schuylkill River, Evans mounted his boat on four large iron wheels and improvised a form of belt drive to the after axle. Thus equipped, the machine majestically waddled down Walnut Street to the riverside and took the water like a gigantic mechanical duck. That done, it was no longer a locomotive but a craft, but thus America took her honourable place with the pioneering nations of mechanical locomotion.

The trouble with the incipient steam railroad at that time was not so much in the infantile ailments of the locomotive as in the hopeless inadequacy of the existing tracks. Iron plateways were really even worse than the ancient wooden-baulk road when it came to supporting a heavy engine and receiving the shocks it gave when in motion. It is not surprising, therefore, that the railway seemed for a while to develop more than the locomotive.

As we have seen, in certain parts of Europe, and certainly in England, mine railways with horse traction were well established. One needs special mention here, for we shall hear about it again in a rather important connection. It was the Middleton Colliery Railway near the rising industrial city of Leeds in the North of England. Its initial claim on history is that, although it was an ancillary undertaking and not a public railway, it was the first line ever to have been built under an Act of Parliament (June 9, 1758, exactly twenty-three years before the birth of George Stephenson).

Two more of these pre-steam lines must be mentioned. At the beginning of the century, England was much troubled by the great Napoleon's Continental System, and railroads were being considered for by-passing the Straits of Dover. On May 21, 1801, the Surrey Iron Railway was incorporated by the

British Parliament, and it was opened from wharfs on the Thames at Wandsworth to Croydon in the South of England on July 26, 1803. This was *the world's first public railway*. It had double track for continuous traffic in both directions; it was laid with flanged iron plates spiked to stone blocks; its traffic was in freight and minerals, and the trains were drawn by horses. Its traffic continued until the coming of modern steam railways, which partly followed its course. An extension called the Croydon, Merstham and Godstone Railway, incorporated in March, 1803, was opened on July 24, 1805. Parts of its course have long been concurrent with the main line from London to the English South Coast, which, electrified since 1933, has long had one of the highest train frequencies in the world apart from underground lines in big cities.

Meanwhile, on June 29, 1804, there was incorporated the Oystermouth Railway Company which opened its line along the coast from Swansea in South Wales in the spring of 1806, though the precise date is not known. This was *the first public railway to carry fare-paying passengers*, who rode in single horse-drawn cars. To later notions it was a tramway rather

A curiosity first published in the 1827 edition of Thomas Tredgold's The Steam Engine. *Modestly presented as a "steam carriage", the design seems to have been Tredgold's own.*

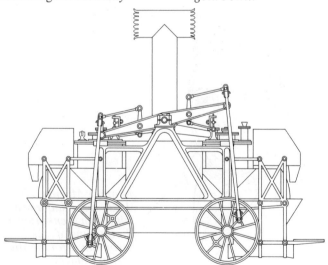

than a railway, and it went on using horse traction for many years. (Steam was introduced in 1877; electric cars in 1929, and the line was at last closed in favour of buses in 1960.)

COLLIERY LOCOMOTIVES

So we come to the first commercial use of steam power on railways, and it brings us back to that old-established Middleton Railway in Yorkshire. It brings us also to a very famous partnership in design, the more remarkable in that at that time, pioneering mechanical engineers usually went alone, as Trevithick, Watt and Newcomen had done.

In spite of the obvious workability of Trevithick's locomotives—apart from their distressing habit of smashing up light iron rails—there was a strong school of thought that for practical purposes, traction through the adhesion of a smooth wheel upon a smooth rail never would succeed. People believed in cog-wheels, which could not slip, and since the oldest example of mechanical engineering known to most was clock-work, that is scarcely surprising. Surely, if one put a Trevithick-type engine, however much improved in power and reliability, to hauling really heavy loads simply by its adhesion to a smooth surface, it would slip hopelessly!

Matthew Murray, 1765 – 1826. He collaborated with John Blenkinsop in the building of rack locomotives for the Middleton Colliery.

24

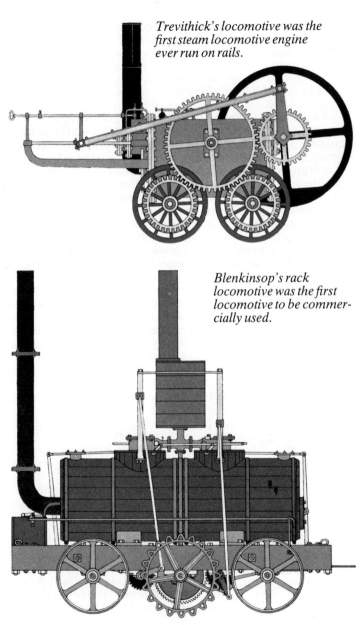

Trevithick's locomotive was the first steam locomotive engine ever run on rails.

Blenkinsop's rack locomotive was the first locomotive to be commercially used.

25

Now in those early eighteen-hundreds, one of the most formidable minds in the improvement of steam engines was Matthew Murray, whose name makes his Scots descent patently obvious. And only a Northern Englishman could have had a name like John Blenkinsop. Murray had invented the short D-shaped slide valve in 1806, greatly improving admission and exhaust events. Blenkinsop's contribution was propulsion by rack and pinion. The rack consisted of a closely regular set of teeth or lugs on the outside of the left-hand rail, and with these engaged a large cog which was the locomotive's driving wheel.

Our representation of the Murray-Blenkinsop locomotive is made from a beautiful model in London's Science Museum, South Kensington. It is fairly self-explanatory, but the following points should be noted. The boiler, with a central furnace leading to a flue at the opposite end, had the two cylinders mounted vertically in the top of its shell. Power was transmitted through transverse cross-head beams to spur wheels both driving the main pinion axle. The exhaust led not to the chimney, as in Trevithick's first locomotive, but to an outlet between the cylinders. Possibilities of making the exhaust produce draught in the firebox had not occurred to Murray. But the open exhaust must have created abominable dins, especially for people who had not been previously accustomed to engines moving about the country. Hence the large wooden silencer which occurs on the model and in the present drawing. Just when this was fitted cannot be accurately recorded; it is lacking in many of the old drawings.

Construction of the first Murray-Blenkinsop locomotive probably began during 1811, and it was certainly in regular traffic at Middleton in 1812, the year in which, be it added, James Fenton produced the spring-loaded safety-valve, a very desirable and important accessory which, however, did not prevent boilers from bursting now and then, usually because of the unsystematic inspection tolerated in early days.

Steam traction is recorded as having been inaugurated on August 12, 1812 with two locomotives named *Salamanca* and *Prince Regent*. This as suggested, was the world's first commercial use of steam haulage by locomotives on rails. Two more locomotives were added in the following year; *Lord Wellington* on August 4 and *Marquis Wellington* on November 23. The names of the engines reflect European power politics of the time, apart from the tribute to His Rather Rascally Royal Highness who later became George IV of the United Kingdom. The general, it will be noted, was not yet a duke, but was rising rapidly in aristocratic status. He had won at Salmanca, but not yet at Waterloo!

Rack-and-pinion propulsion was to continue on the Middleton Railway until 1835, by which time steam lines of orthodox sort were entirely established and appearing all over Europe and in many parts of North America. The principle then became dormant until it was revived, in less primitive form for steep-grade mountain railways, as on Mount Washington in the United States and very soon after on the Rigi in Switzerland.

The Middleton trains were scarcely rapid things. The famous English engineer David Joy saw one in infancy. Having been told that it would come by "like a flash of lightning", the little boy was somewhat disappointed that "it only came lumbering on like a cart".

Still in England, our scene shifts now from Yorkshire to the real North-East, to that Northumbria which once was a kingdom in its own right, and more recently had become the stage for railway development. Wylam Colliery, near Newcastle, had had a wooden railway for many years. Beside it, George Stephenson had been born. But it was not he, but a partnership again, that brought locomotives to Wylam. In spite

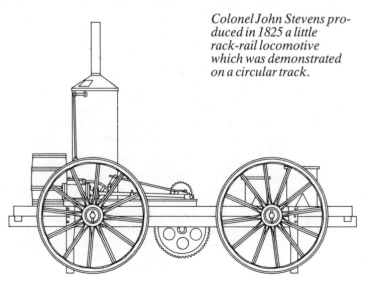

Colonel John Stevens produced in 1825 a little rack-rail locomotive which was demonstrated on a circular track.

of Blenkinsop's rack and pinion, faith in the smooth wheel on the smooth rail did not die. After all, Trevithick had shown that locomotives *could* run by adhesion. William Hedley at Wylam was of like persuasion, and for him in 1813, Christopher Blackett built some very famous old locomotives. No high-sounding names from the Napoleonic wars here! The local people called them *Puffing Billy* and *Wylam Dilly*, and the names stuck. There was a third, nearly forgotten one, named *Lady Mary*.

The first two engines are still in existence, treasured relics in London and Edinburgh. That in London is generally assumed to be "Billy" while the one in Scotland is "Dilly". There is a splendid replica of the former in Munich, built many years later by the Bavarian State Railway Works. It must be said straight-away that these relics do not show the engines in their original condition. As built, they had flangeless wheels and ran on flanged cast-iron rails which they inevitably smashed to pieces, as Trevithick's engines had done before.

To distribute the weight more easily (even empty, each engine weighed about seven tons) new frames were made, giving the engines eight wheels instead of four, the axles being grouped in pairs of supplementary frames resembling bogies, though flexibility on curves was not the object in this case.

In later years, both engines were altered back to the original two-axle arrangement, but with flanged iron wheels to run on iron edge-rails. It was not until 1820 (October 23) that a patent was granted to John Birkinshaw of Bedlington Iron Works, in North-eastern England, for the making of rolled malleable iron rails. Only then did the railroad—as we understand the term, meaning a complex system of heavy transportation at reasonably high speeds—become possible. The term "railroad", indeed, is the old English word for such a system, distinguishing it from the much lighter and more primitive "railway". Unfortunately, it was dropped in British usage though America most happily perpetuated this splendid, sonorous word.

To revert to "Billy" and "Dilly". They had boilers of wrought-iron plates, each containing furnace and return flue so that firedoor and chimney were at the same end, as in the original Trevithick type. (Generations of museum cats in London's Science Museum at South Kensington have had their kittens in *Puffing Billy's* inaccessible stomach!) The cylinders were vertical, as in Matthew Murray's Middleton engines, but they were outside, a progressive step long ignored by other pioneer designers. They exhausted into the chimney.

Transmission was by beams and spur-wheels, and the valve gear was overhead. A four-wheel tender carried coal and water.

What a distant world was that of 1813! Great Britain and the United States were having a second, and futile, war. Napoleon still straddled much of Europe, but had lost the adoration of Beethoven, who had yet to compose his Ninth Symphony. Up in the extreme North of England, *Puffing Billy* was moving coal, except when derailed, which was not seldom.

At that same time, George Stephenson, son of a colliery fireman, was Enginewright at Killingworth Colliery, while the smith at Wylam, Stephenson's birthplace, was Timothy Hackworth, of whom also much more was to be heard.

Puffing Billy *was designed and built by Christopher Blackett and William Hedley, in 1814.*

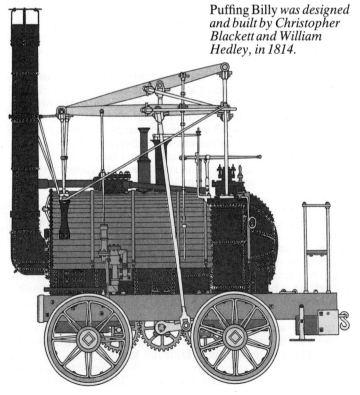

Stephenson was self-educated in a very hard world, that of poverty. Hackworth was an intellectual mechanic of the period, a worker for six long days and a lay preacher on Sundays.

Stephenson moved into locomotive work. Not an entirely original inventor, he was a great improver, a perseverer, and a shrewd business man; which qualities made him one of the historic immortals of applied technology. In his early locomotives he copied Murray's arrangement of boiler and overhead cylinders with their motion and gear, but he would have nothing to do with rack-and-pinion propulsion. It was adhesion from the first. In 1814, he built for Killingworth Colliery *My Lord* (named for the proprietor, Lord Ravensworth) whose two cylinders drove the two axles through gears and counter-shafts. Next year, he discarded this arrangement for direct drive, with the axles coupled by chain and sprockets, and in 1816, he was using a valve gear worked by loose excentric. The chain coupling was apt to break and was very noisy. It was fairly soon to be succeeded by coupling the wheels by side rods and cranks, which would persist throughout the history of the orthodox steam locomotive and in the design of earlier electric locomotives also.

Various Stephenson locomotives were built for industrial lines in the English North-East over the next ten years, though cable-haulage with stationary engines came into considerable use, and was indeed essential on any appreciable gradient: George Stephenson had been given charge of such an engine when he married, as far back as 1802 when he was twenty-one. Two very old Stephensonian locomotives lasted in service for an immense time, one for some eighty years, and have survived, one at Newcastle and the longest-lasting, from Hetton Colliery, at York.

Suspension was an important matter, especially with the uneven tracks of the time. All wheels had to be sprung and there was a short phase in which the springs consisted of plungers contained by cylindrical "legs" in the bottom of the boiler shell, open to the water space, so that the actual boiler pressure provided the necessary resilience. When not in steam, the engine sat down on its axles. It was an ingenious but fallible arrangement.

Stephenson of course was not alone. Hackworth, as suggested, was in the running, with honest, dogged, and sometimes ingenious work. Some others designed, and even built, locomotives of more fantastic sort. Perhaps the most absurd was Brunton's, which rested on two free axles and was expected to push itself and its load along by mechanical legs in the rear. It did succeed in exploding its own boiler, with bloody results for some people.

That term "steam carriage" reminds us that in these years there was intense interest in steam propulsion, *especially for passengers*, on ordinary main roads; indeed the steam motor-car very nearly arrived with the train. It is one of the tragedies of mechanical history that the two did not develop simultaneously on a proper commercial scale. One of the steam-carriage pioneers, Sir Goldsworthy Gurney, must be mentioned in our present connection. In 1826, he produced three most important inventions; a firebox fusible plug which, by melting, flooded the furnace with steam if the boiler water-level fell dangerously; a valve gear designed to use steam expansively instead of by a primitive sequence of admission and exhaust; and a multiple-jet blast-pipe to give even draught from the engine exhaust turned into the chimney.

THE STOCKTON AND DARLINGTON RAILWAY

George Stephenson was now in his forties. His son Robert was grown up, and through some toil and privation had been given the advanced technical education his father had never enjoyed. The inevitable European depression after the Napoleonic Wars was righting itself in heavy industrial

In the early days of the railway, signals were given by simply waving the arms. On the left, the stop signal, on the right, the warning signal.

expansion; in England, France, and even in feudal Germany and impoverished Scotland. The last-named had tried steam on its Kilmarnock and Troon Railway as far back as 1817, with the then usual results: cast-iron rails smashed to atoms and the engine discarded as a destructive machine. The Germans had tried, and even built for themselves in the Royal Foundry, Berlin, two Murray-Blenkinsop-type rack-rail locomotives with tragi-comic results.

But, as noted, Birkinshaw's rolled iron rails had arrived, and they were to work a great magic. The River Tees was the natural outlet of the Durham coalfield in the English North East. Between Darlington inland, and Stockton on Tees, there was need for a railway on a more ambitious scale than before. The great Quaker family of The Stockton and Darlington Railway was incorporated in 1821. Its Act of Incorporation

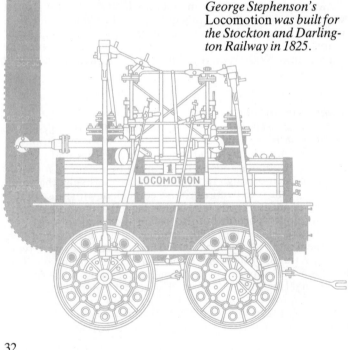

George Stephenson's Locomotion *was built for the Stockton and Darlington Railway in 1825.*

George Stephenson, 1781 – 1848. Almost entirely self-taught, he rose from being engine-wright at the Killingworth Colliery to become the leading locomotive and railway engineer of the first half of the 19th century.

allowed for its working by "men, horses or otherwise". Stephenson suggested (probably with some force, being himself) that "otherwise" could and ought to mean "by steam locomotion". Edward Pease was cautious, as Quaker people are, with much silent comment after their manner, but he was a man of his word, after their kind. The end of it was that the Stephensons, father and son now, produced the first steam locomotive to work public traffic regularly on a company-owned right of way.

It will be seen that *Locomotion* was still of the primeval Stephenson type, with her cylinders and "works" on top, the former driving cranks set at ninety degrees front and rear on each side. Coupling rods were arranged by means of a return crank on each driving crankpin on the rear wheels, the front bearings of the rods being of course coincident with those of the forward connecting rods. The boiler, lagged with wooden strips over its upper portion both to retain heat and to spare the enginemen, had the old medial and furnace-and-flue arrangement, with no smokebox. The wheel centres were built up of perforated iron segments with iron tyres shrunk on. The tender was simple. On a frame resembling that of the traditional English *Chaldron* coal wagons it supported a cistern for the feedwater, with space for fuel below.

Late in 1824, the Stephensons' *Locomotion* was under construction in their Forth Street workshops at Newcastle-upon-Tyne. She was ready in the following year, but ere the railway was ready for opening, the young Robert Stephenson was far away on a mining survey in northern South America. (There, incidentally, he was to find the impoverished Richard Trevithick, and to help him home.) The final irony may be added here; for Trevithick was to die in England, poor and distressed in 1833, having often petitioned the British Government for some sort of civil pension, which was always refused. Robert Stephenson (and again without Government assistance) was to become the first millionaire engineer (sterling, in gold) and, when he died at the age of fifty-six, he was buried with the kings and the great soldiers in Westminster Abbey. Quaint!

Now, there was another thing about this Stockton and Darlington Railway; it was to take passengers as well as goods traffic. With this in view, a solitary passenger coach was built and included in the make-up when the inaugural train, headed by *Locomotion*, ran over the line on September 27, 1825. Its exact form cannot be reproduced. For many years, there was current a drawing of a sort of shed-on-wheels, but it seems to have been an ingenious adaptation of the road coach of the period, judged by J. R. Brown's beautiful pencil drawing, recently discovered and believed to have been made at the time.

It was marshalled in the middle of a train of coal wagons, with the important personages sitting in it while as many as

could climb on to the wagons did so. Crowds had come to see the prodigy, some at least because they thought *Locomotion* would explode and thus give them a nice kick. But all went off without a hitch. All the same, in regular traffic, for some time passengers were conveyed by horse-drawn rail coach in between the steam coal trains. People were still rather nervous about riding behind a ferocious, fire-breathing machine which, they had been credibly told, could blow both itself and them to bits if it suffered from mechanical indigestion.

Still, the thing was done. The steam railroad, which previously had been regarded as a sort of industrial machinery, as we today regard mining equipment and telpherage, was at last present as a means of mechanical public conveyance. It was not an entirely happy occasion. George Stephenson was still without his son Robert, who had done so much with him in the improvement of the engine. Edward Pease, the faithful backer and sponsor, had just lost a son, and was full of the sorrow that he had been trained never to show. Still the majority of the British aristocracy had little awareness of what was going on, though later it was to improve its fortunes quite considerably.

The revolution (and, thinking of France, they did not quite recognize the term in this connection) had really begun.

Revolution indeed was a word conveying different meanings to different people. To the French it meant Reform. To the Americans it meant Glorious Independence. To the English landed aristocracy it meant social upheaval that none of them wanted and, at worst, all the excesses of the French Terror under Robespierre. They wished no revolution, mechanical or political, and believed that the one would lead to the other. Where railways might adventure out into the country from the industrial districts, wherein they might be tolerated as a necessary evil, all schemes were fiercely opposed, both by political pressure and by actual violence against parties of surveyors. The great northern commercial and industrial families, on what was then the political Left, were of course greatly in favour of this magnificent invention, and that is how, during the eighteen-twenties and early 'thirties, the thing became a party issue in Parliament.

EARLY DESIGNS

Visionary Englishmen like Thomas Gray and William James foresaw, and backed even to bankruptcy, a national railway network with steam traction. But ere that could happen, not

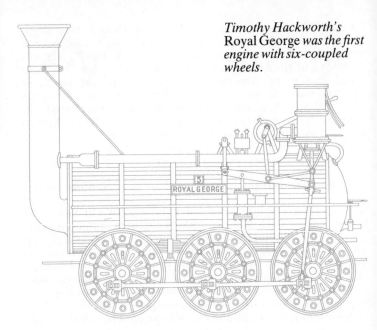

Timothy Hackworth's Royal George *was the first engine with six-coupled wheels.*

only must government be influenced, but great improvements were still necessary in motive power itself, before it could be adequate for scheduled passenger and freight movements. At this stage, George Stephenson was passing rather out of the locomotive picture into that of Civil Engineering. In the late 'twenties, he was busy with the survey and construction of a real trunk railroad between cities, the historic Liverpool and Manchester Railway. Robert Stephenson was home again (with the tragic Trevithick) but now two very important men enter our stage.

One was the sober-sided Timothy Hackworth, who had taken charge of the Stockton and Darlington Railway's engine-works at Shildon. The other was Marc Séguin in France. Hackworth was the more practical and Séguin the more academic, as one might have expected with a puritan Englishman and an intellectual Frenchman. Let us take Hackworth first.

Most important of his work at this time was the locomotive *Royal George*, which he built for the Stockton and Darlington Railway in 1827. Its origin was odd. Previously, a man named Robert Wilson of Gateshead in the same county had built an

Timothy Hackworth

Marc Séguin

engine with a long boiler and four cylinders, two to each axle. Desire outran performance, though the idea was bold. The Stockton and Darlington company bought the engine simply for the sake of its large return-flue boiler. This Hackworth mounted on a six-wheel frame with two vertical cylinders high up, driving the same cranks as those which served the side coupling rods. He had previously made a model to see how this six-coupled, direct-drive arrangement answered. The model had worked well. So did the *Royal George*. Though slow, she

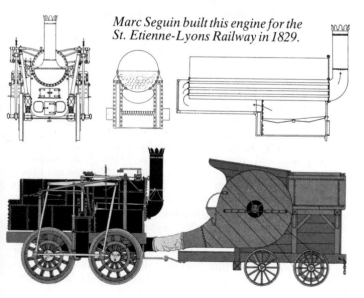

Marc Seguin built this engine for the St. Etienne-Lyons Railway in 1829.

was a heavy puller such as none had known before, admirable for Durham coal traffic. In the formulae of much later years, her wheel arrangement was describable as 0-6-0 (Anglo-American), 0-3-0 (French) or C (German, in which an alphabetical sequence was and still is used for counting powered axles).

Railways for coal haulage came to Central France at this time. Beaunier's horse railway from St. Etienne to Andrézieux had come in May, 1827, and in 1829 trials were being made with steam on the Givors to Rive-de-Geir section of the St. Etienne and Lyons Railway. For this, Marc Séguin had already obtained some rather odd engines from the Stephensons, but in 1828, he patented a multitubular (fire-tube) boiler which he had incorporated in a locomotive late in the following year. Models have survived. It was altogether a more advanced boiler than anything in previous Stephenson or Hackworth practice. The draught arrangements, however, were clumsy, consisting of immense rotary fans mounted on a tender and feeding air to the furnace through bellows-like leather pipes.

Marc Séguin, unlike his British contemporaries, was rather a practical scientist than an advanced enginewright. Ingenuity ran in the family. Joseph and Etienne Montgolfier, the pioneers of ballooning and makers of the first airborne craft, were his uncles. The idea he had incorporated, that of providing heating surface by a large number of small flues passing through the water in the boiler, had been patented by James Neville in England, in March, 1826. Marc Séguin and his brother were engaged in many notable projects, including early steam navigation on the Rhône.

Before going on to steam power's next applications we must notice some other landmarks. The first railway in the old Austro-Hungarian Empire, generally called Linz-Budweis, was interesting in that from the beginning it was a passenger line, but only horse traction was used when its first section was opened from Budweis (Ceské Budejovice) to Trojanov in what is now Czechoslovakia, on September 7, 1827. It was chiefly the work of Franz Zola, father of the Emile Zola of French literature.

The advent of steam locomotion on rails in the United States was tragi-comic; a pity, seeing how its use was to make possible the ocean-to-ocean form of the great republic. Important canals had already been built in the Eastern States, and in the summer of 1829, the Delaware and Hudson Canal Company obtained a steam locomotive from Foster, Rastrick and Company in England, for use on an auxiliary railway which it was building

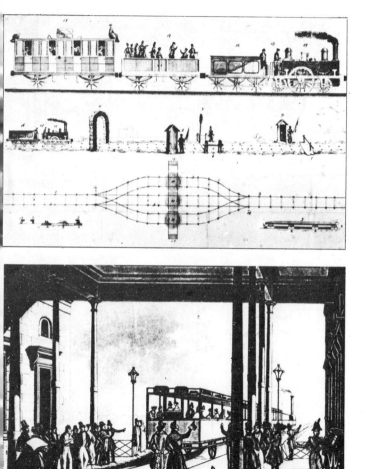

39

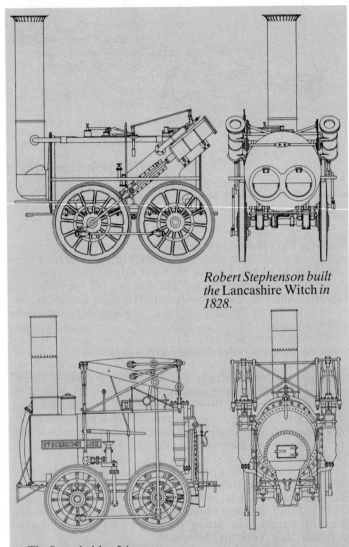

Robert Stephenson built the Lancashire Witch *in 1828.*

The Stourbridge Lion *was the first actual locomotive to run in America.*

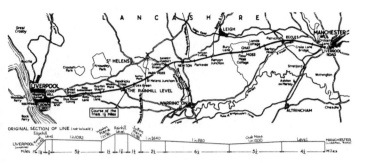

This map and gradient profile of the Liverpool and Manchester Railway shows where the Rainhill Trials took place.

between Carbondale and Honesdale. The engine was named *Lion*, but since there were many *Lions* in the course of time, and in view of her English Midland origin, she is generally called the "Stourbridge Lion". She was one of four ordered by Horatio Allen of the Delaware and Hudson Company; two more from Foster Rastrick, and one, of which more later, from Robert Stephenson.

Allen had been to England to see what was going on there, when he made the order. On her arrival at Honesdale, *Lion* was deemed too heavy for service. True, the road was laid with quite respectable iron rails, but people were anxious about a wooden trestle over Lackawaxen Creek. To justify his purchase, Mr. Allen drove her over the trestle and back, without any fearsome results, but the Board was not convinced, and the unfortunate engine was laid by, eventually to be broken up. The other two Rastricks and the Stephenson never reached Honesdale. Some limbo in New York claimed them.

On October 9, the railway inclines and levels of the canal were opened, but they were worked by cables and gravity. None could guess that, as well as coal, some day the Delaware and Hudson Company would be handling not only millions of tons of freight, but lifting passengers on overnight plush between New York and Montreal.

That the "Stourbridge Lion", as an engine, was quite all-right, even though by then she was slightly archaic, is suggested not only by Horatio Allen's doughty championship of her on that trestle, but by the record of an almost identical

41

engine called *Agenoria* which never left England but worked usefully for some thirty years on the Earl of Dudley's colliery line, the Shutt End Railway, in her native West Midlands. Some bits of the *Lion* were later unearthed, and *Agenoria* has long been preserved. Horatio Allen's brave but sad adventure with the *Lion* ended his very important results. At the time, however, his prospects may well have seemed hazardous.

There was this difference from a similar situation in Europe. America received any new machine, if it were seen to be practical and likely to further the interests of a new and rapidly growing nation, with delighted enthusiasm. In Europe, substantially powerful classes viewed it with distrust, alarm and even dismay.

We must return awhile to England. A rather important type of locomotive was just being essayed by the Stephenson firm. It still ran on four wheels, all coupled. But it had inclined outside cylinders, instead of vertical ones whether outside or embedded in the boiler-top as in the ancient Murray form. One such was the *Lancashire Witch* built in 1828 for the freight-hauling Bolton and Leigh Railway. The boiler was still old-fashioned, though instead of a single or return flue, it had two furnace-flues side by side, leading into a common chimney. Akin to the "Witch" was the *America*, the engine which Stephenson built for the Delaware and Hudson Company, but which never got nearer to it than New York. A third was called *Invicta*, which went into service in the South of England on the Canterbury and Whitstable Railway, opened on May 3, 1830, and for which may be claimed that it was the world's first line to carry both freight and passengers entirely by steam power. But the little locomotive was a relatively feeble thing, confined to one short level section of the single-track line, stationary engines and cables doing the rest of the work.

It was between Liverpool and Manchester in the English North-West that the world was first to see a railroad as succeeding generations would understand it, with all sorts of traffic hauled by locomotives, on two separate roads for opposite directions, with real stations, proper schedules, and trains of reasonably respectable rolling-stock. Behind the enterprise was a group of rich Lancashire capitalists who later became so powerful that the business world called them simply *The Liverpool Party*.

They had a long and bitter fight. Several lines were surveyed before that of George Stephenson was adopted, with its imposing rock cutting and tunnelling in Liverpool, and its magnificent multi-arch viaduct across the Sankey Valley. Much could be written about its vicissitudes from the time of its incorporation on May 15, 1826; of landowner-trouble, of the way in which the elder Stephenson laid his solid road across Chat Moss, a moor which had been regarded from time immemorial as a bottomless bog wherein a horse and his rider could be engulfed, to vanish for ever. This, however, is a chronicle of the train, not of railroad building.

With the line at length authorised, the company showed itself suddenly shy of locomotive traction, favouring cable haulage by stationary engines. George Stephenson, who had a ferocious tongue when roused, fluent with four-letter and other words, demanded that they should recognize the ability of the conqueror of Chat Moss to provide traction. He had been building locomotives for over a decade now.

The directors so far relented as to offer a prize of £500 sterling for a locomotive which should fulfil certain very stringent conditions as to weight, power and speed. Nor did they simply challenge Stephenson to produce such an engine. They made it an open competition, and whoever produced the most satisfactory locomotive would clearly land a contract worth a fortune.

Watt's "sun-and-planet" drive as proposed for a Stephenson locomotive, c. 1828.

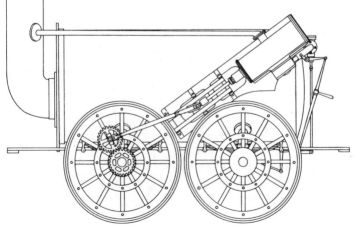

THE RAINHILL TRIALS

There were plenty of competitors, some of them, indeed, from the lunatic fringe of their profession. The serious ones were George and Robert Stephenson in partnership with Henry Booth; Timothy Hackworth on his dogged own; the ingenious Swede Captain John Ericsson in a rather unfortunate partnership with the Englishman John Braithwaite; Timothy Burstall, of Leith in Scotland, who had designed a steam road carriage (an aversion of George Stephenson, be it remarked); and Edward Bury of Liverpool.

Timothy Hackworth's
Sans Pareil

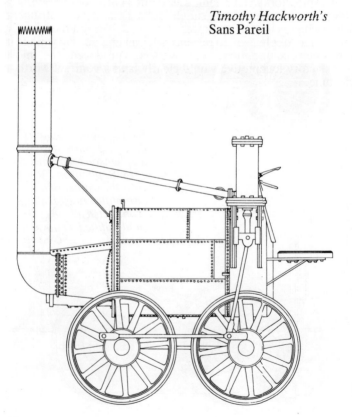

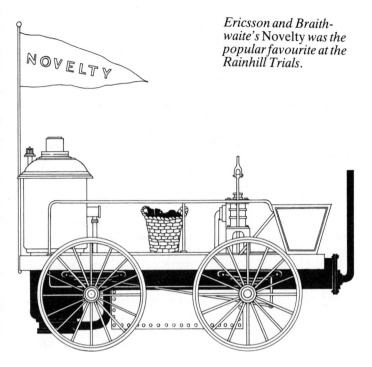

Ericsson and Braith-waite's Novelty *was the popular favourite at the Rainhill Trials.*

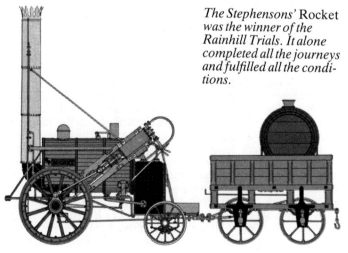

The Stephensons' Rocket *was the winner of the Rainhill Trials. It alone completed all the journeys and fulfilled all the conditions.*

The last-named, though later of some eminence in locomotive engineering, could not produce an engine in time for the trials, which were held on the Rainhill Level of the Liverpool and Manchester line, from October 6 to October 14, 1829.

Getting the engines from factory to Rainhill presented plenty of problems. Burstall's engine from Scotland was damaged in transit, though it is reported to have made a short run or two. It was an effete little thing with a small upright boiler between large wheels, an adaptation of its designer's steam-carriage engine.

Hackworth's engine, *Sans Pareil*, was perhaps describable as a shortened, four-wheeled version of his *Royal George*. She did not strictly fulfil the conditions as to the weight allowable on two axles only, but she certainly *went*, if rather ponderously, consuming an inordinate amount of fuel. One of the conditions was that the engines *should effectively consume their own smoke*. All the entrants got round that one by firing with coke instead of coal. But the unfortunate *Sans Pareil* had bad luck apart from her astonishing appetite. She broke down on the eighth trip over the course; her water-level fell and she dropped her fusible plug, filling the firebox with steam. For what it is worth one must in fairness remark that her cylinders had been cast by Stephensons, and were not faultless.

That left Ericsson's *Novelty* and the Stephensons' *Rocket*, with the former a hot favourite.

Captain John Ericsson, sometime of the Swedish Army, was then twenty-six years old, and not yet a legendary figure. Those who liked him least regarded him as a mechanically minded mountebank, and there were to be long and bitter years before he was to achieve worldwide fame and fortune in the United States with the *Monitor* warship. He had come to London in 1826, and had met that lesser figure, John Braithwaite, who in later years was to be a singularly inept engineer of the Eastern Counties Railway. The locomotive *Novelty* bore all the marks of Ericsson's genius and she was to suffer from the failings of ingenuity without practical trial. She was a four-wheeler with vertical cylinders, most beautifully designed and made, a prototype for a light locomotive which really would have been much better suited to road transport. She carried her fuel and water-supply on the main frames, the latter in a big tank below and a smaller one aft. Her weak points were in her boiler.

Now this was a very ingenious boiler. But it had serious weaknesses, as we shall see. Its main portion was vertical, with a copper casing, fired through a funnel in the top. From this, a much narrower portion went horizontally to the rear, so that the whole steam-generator was something like a T laid on one side. From the furnace, a back-and-forth flue led ultimately to a tall chimney in the rear. It was a clever way of obtaining heating surface; also the greater proportion of this was in the firebox, an important virtue, as later experience was to show. It was the engine's light and very elegant appearance which chiefly endeared her to the spectators. When she skimmed along at thirty miles an hour their enthusiasm was intense. Draught was induced by bellows, always a weak point, so the sharp puffing of Hackworth's and the Stephensons' engines was absent; she went more like a steam motor-car, which was ever a pleasantly silent thing, and that pleased the crowd still more.

Indeed, the alarming aspect of a steam locomotive had been much argued by its angry opponents. During one of his constant catechisms in Parliamentary Committee and elsewhere, George Stephenson had had it suggested that the spectacle of a locomotive engine working hard, with a red-hot chimney, would terrify cattle, with disastrous results of all sorts.

Now old George, as a boy, had once herded cows, and knew something about them. He said that the cow was not the most intelligent of beasts. The cows would doubtless think that the

(Over) *Painting by Bouw-meester of an old steam locomotive in the woods near Arnhem, Holland.*

This engraving of the Novelty—*probably with Captain John Ericsson driving—is by Vignoles.*

engine had merely had its chimney painted red, and would not bother themselves further. But perhaps to show that a properly-driven locomotive never would have a red-hot chimney, he painted the tall stack of the *Rocket* white.

The *Rocket*, one believes, was largely Robert Stephenson's work. She had a multi-tubular boiler like Séguin's, fired by a water-jacketed firebox at the rear, believed to have been the work of Henry Booth, the Stephensons' co-entrant in the competition. She had the inclined outside cylinders of the *Lancashire Witch*, *America* and *Invicta*. As a curiosity, by the way, we give a tracing from an ancient Robert Stephenson drawing which shows such an engine with James Watt's "sun-and-planet" drive instead of the by-then-normal cranks.

The *Rocket* needs little further introduction and her beautiful simplicity is evident from the drawing. The colours are authentic. The tender (still with a water-barrel instead of a proper tank) was built by Nathaniel Worsdell, a noted English coachbuilder of the time. After many vicissitudes, the engine's worn-out shell still exists in the Science Museum, London. There are several full-size replicas, one in the same place, sectioned for display, and more in America, the first of which was commissioned from Robert Stephenson and Company by Henry Ford. We have seen the Ford engine working under steam.

Now for the Rainhill contest! Each engine was required to run seventy miles continuously (that was in fact back-and-forth over the relatively short course) at not less than an average speed of ten miles an hour. *Novelty*, as we have seen, managed a *maximum speed* of thirty miles an hour. ("It seemed to fly..." wrote one spectator, and it must be remembered that never had anybody seen anything going so fast, apart from eagles, swallows, ducks and some other birds!)

Other conditions must be mentioned. The accepted engine must not weigh more than six tons on six wheels, or 4.5 tons on four wheels; its working pressure must not be more than fifty pounds a.d.p. per sq. in., its boiler having been previously subjected to a water test equal to thrice that pressure; it must have two safety-valves and a mercurial pressure gauge; it must be able to pull a twenty-ton train at ten miles an hour. The *Rocket* and the *Novelty* both fulfilled conditions, but the latter's ingenious yet dicey boiler was her undoing. There were two failures, and then she was withdrawn; collapsed though not indeed exploded.

So poor little *Novelty*, as well as Hackworth's massive *Sans Pareil*, was out of the contest, leaving only the Stephenson

engine *Rocket*. At that, one wonders less that some parties had brought forward machines without benefit of artificial power. Brandreth's *Cyclopede*, for example, was worked by a horse trotting on something like an escalator or *trottoir roulant* connected by gears to the axle below, and there was even a "manu-motive" approximating to the hand-car of later years.

But the *Rocket* stayed the course, completing all her trips, loaded, without any accident. To the Stephensons and to Booth went the £500 prize, and to the former went the contract for supplying locomotive power to the Liverpool and Manchester Railway. The *Rocket*'s maximum speed was twenty-nine miles an hour, and her average was sixteen. She was sold to the company for another £500, the stipulated maximum price having been £550. Up in Newcastle-upon-Tyne, eight more "Rockets" were put in hand for the opening of the railway in the following year. Improvements were made from engine to engine. Firstly, and in the *Rocket* herself before she went into service, the angle of the inclined cylinders was greatly decreased. The original high angle had resulted in very alarming oscillation. Then the boilers were given proper smokeboxes. The blast-pipes in these were much improved to create adequate vacuum at the ends of the flues. The engine *Northumbrian* (August, 1830) had a much advanced firebox, completely within the boiler barrel as to its upper part, and she also had a proper tender instead of a coal-truck with a water-barrel on top.

Of the other contestants, Hackworth's *Sans Pareil* was found useful, and still survives at the Science Museum in London, in her original state as far as one can tell. Braithwaite and Ericsson made two more locomotives of "Novelty" type, but larger, and with fan drive instead of bellows, which were to work on the St. Helens and Runcorn Gap Railway. *Novelty*'s wheels and cylinders were unearthed many years later. There is a handsome replica, containing original parts, permanently exhibited beside her two rivals.

Thus was answered the question of how this Liverpool and Manchester Railway, the first inter-city main line for passengers and freight, should be worked. It was to be done by locomotive, with no more nonsense about stationary engines and cables. The locomotive, as a machine in constant public service, had *arrived*. It is wrong to call George Stephenson the *Father of the Locomotive*. He was not; though many people were to call him so—quite angrily when disputed—for many years. But one may safely remember him as the *Father of the Steam Passenger Train*, indeed of all trains as we have known them. When he was old, he told a young man that a time would come when electricity would be the great motive power of the world. He would not see it, he said, but the young man might. Be it added that at that time the power of electricity was generally unknown, or at most regarded as a sort of laboratory magic. Alessandro Volta (1745–1827) was fresh in the memory; Michael Faraday (1791–1867) was in his prime. To most people who had ever heard of him outside American revolutionary politics, Benjamin Franklin (1706–1790) was a crackpot character who had flown a kite to catch the lightning! For the present, it was Steam's Day. For a long time, steam had been pumping water out of mines, then hauling mine-shaft cases, tip-wagons, and working machines in mills. Now it was shown able to move freight and passengers with some speed over apparently illimitable distances. To some, those speeds of twenty to thirty miles an hour were against Nature and therefore quite blasphemous. With this new machine, mankind was clearly infected with the madness of the Gadarene swine, and thus doomed to a miserable end in double-quick time. In England that old general, the Duke of Wellington, feared that such rapid mechanical progress would enable fierce revolutionary mobs to capture the country, while less politically-minded critics foresaw a dangerous ability to circulate among the criminal classes (a situation which was indeed to come to pass, over a century later, with the advent of the cheap motor-car!). The general, by now a rather unsuccessful Prime Minister, need not have worried so much. A revolutionary mob *might* capture a train, but it was even more easy for Government to requisition a few, load them up with soldiers, and rush them to wherever they might expect trouble.

With that irony which is so peculiarly English, the Duke of Wellington formally opened the Liverpool and Manchester Railway, on September 15, 1830.

Really, it was a most inauspicious function! For one thing, Wellington was cordially detested by the mass of the people in North-Western England, which was extremely Radical. Then the weather was fearsome, with terrific thunder showers. During a leg-stretching halt at Parkside, on the way from Liverpool to Manchester the *Rocket* ran over William Huskisson, sometime President of the Board of Trade, who had lately quarrelled with Wellington, and crushed his legs so that he died the same night. Festivities were abandoned, but Wellington proceeded to Manchester in the ducal train, and was there received with screams of hatred and even brickbats.

Late in the evening came what may have been the world's

first serious attempt at train wrecking. Some angry young man—or it may have been even a squire's undergardener—put a wheelbarrow on the line as the trains returned. The wheelbarrow was damaged beyond repair, but the trains continued to run.

Such was the beginning of public, power-hauled, passenger and freight railroads which were to cover much of the world's land masses before the century was done. The rail was to have its tragedies, and many of them, but its history was to be less sinister and less gory than those of other agents which were to come later. It was left to the automobile to spawn the tank; to the aeroplane to produce the bombing aircraft. The ghosts of those mild and peace-loving Quakers, the Pease family, might well console themselves.

Chapter 3
The Great Morning of Steam

In the United States of America we left that remarkable man Horatio Allen quitting the little auxiliary railway of the Delaware and Hudson Company, disappointed and by some discredited. But he had that perseverance which George Stephenson had, facing similar adversity. America was very different from England, as a country; indeed, a common language has often led to trouble between decent people who did not realize this.

In the United States, then a much smaller country than now, but still very large by any Western standards save the Russian, there was far less distrust of the new engines than in England or in most other European countries. The British ruling classes often saw in the engine something that was to upset their way of life and turn the national house out-of-doors. The immediate commercial success of the Liverpool and Manchester Railway, whose directors were themselves astonished by the freight as well as the passenger volume, merely exasperated the landowning class. But the Americans, striving to turn strips of coastal or riverside ex-colonies into a mighty nation, saw in the infant railway train a tremendous peaceful weapon which could do just that, and they received the steam railroad with delight.

To be sure, there was to be plenty of real fighting in the States thereafter, and not yet was the time when trains would run (in relays) from Atlantic to Pacific. But the great waters could be linked; even the Atlantic with the Mississippi Basin. It was not easy. By European notions the country was incredibly wild. There was little native mechanical engineering; a village blacksmith was quite a personage, cf. Henry Wadsworth Longfellow, one of the greatest of earlier American poets, and an enlightened man in several other ways. The sole seat of mechanical knowledge was in the engineering faculty of the United States Military Academy, West Point, New York State.

On February 28, 1827, the State of Maryland granted a Charter to the Promoters of what was to be called the Baltimore and Ohio Railroad. In the survey, the United States Army greatly helped. The original route included the splendid Carollton Viaduct, built of Maryland stone and begun in 1828. The foundation stone of this was laid by Charles Caroll who had given his name to the place and was even then the last survivor of those who had signed the American Declaration of Independence. By half a decade this remarkable structure anticipated the much-more trumpeted Thomas Viaduct at Relay, Maryland, which was completed in 1835. Between them, among great arched railway structures came the Sankey Viaduct in England (1829–30), though none of them came near, in antiquity, to the Causey Arch in Northumberland, which was out of useful service before they were built.

But back to *The Train*! By the beginning of 1830, the Baltimore and Ohio Railroad (which was not to reach the Ohio until 1863) mustered fourteen miles of route on double track

Peter Cooper's Tom Thumb *in its famous race with the horse-drawn car.*

from Baltimore to Ellicot's Mills, worked by horse-cars, on wooden road with iron straps a-top.

Peter Cooper, a rich and ingenious New Yorker of the period, made a very small experimental locomotive, the *Tom Thumb*, for demonstration on this new railway. She had a vertical multitubular boiler with its flues made from sawn-off musket barrels, geared drive and draught induced by belt-driven fans. In 1830, with benefit of double track, Cooper raced *Tom Thumb* against one of the horse cars. The great horse went off at a fast trot. Then the little engine overhauled him. But the belt driving the fan began to slip. Cooper tried hard and heroically to keep belt on wheel, but only got his hand sawn, as one does when trying tricks with rapidly moving belts. So the Horse won that race. But still Peter Cooper had shown some people something that otherwise they would not have believed.

More heroic notions were being realized in the South. What was to become the South Carolina Railroad was surveyed to connect the Atlantic Coast at Charleston with the township of Hamburg, on the banks of the Savannah River opposite the modest but rising city of Augusta, Georgia.

Thither went Horatio Allen, recovering, like a good American, from certain disappointments with the Delaware and Hudson Company in the North. He was made Chief Engineer of the new railroad in the second half of 1829, about the time the Stephensons, Hackworth, Ericsson and lesser people were busy convincing cautious Lancashire business people about the desirability of steam locomotives, up in the English North-West. Allen in his turn, early next year, convinced his directors that steam should be used exclusively. He had an able mechanical engineer in E. L. Miller of

The Best Friend of Charleston *was the first locomotive in America to pull a train.*

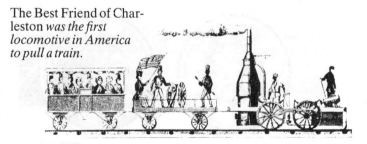

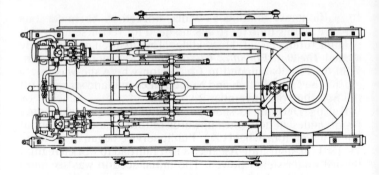

Best Friend of Charleston, *the first All-American steam railway locomotive to go into commercial service, 1830.*

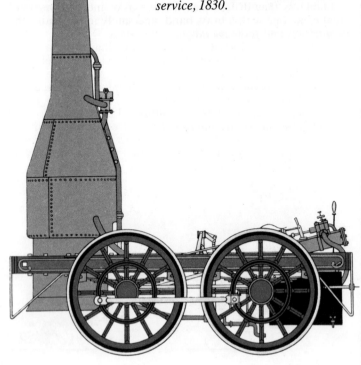

Charleston. West Point Foundry undertook to build four engines, so this time there was no need for recourse to British manufacturers, as at Honesdale. The first engine, happily named *The Best Friend of Charleston*, was ordered in March, 1830, was safely shipped to Charleston where she was safely erected by Julius Petsch and Nicholas Darrell, and steamed for the first time on November 2.

During December, she ran several trips, just exceeding twenty miles an hour with upwards of forty passengers (about ten to a light four-wheel car). Public service began on January 15, 1831, inaugurating America's first railroad to work a regular service under steam power. The initial section was six miles long, out of the pleasant city of Charleston. Though the present venture was more modest, as yet, than the previous one between Liverpool and Manchester in England, the event was certainly happier and nobody was damaged, not even an ex-Cabinet Minister. The first of the little cars carried soldiers with a very light field gun, with which they joyously fired rounds of blanks as they drifted past the live-oaks and cottonwoods. Next came one with a brass band, and safely in rear came the proprietors and their friends.

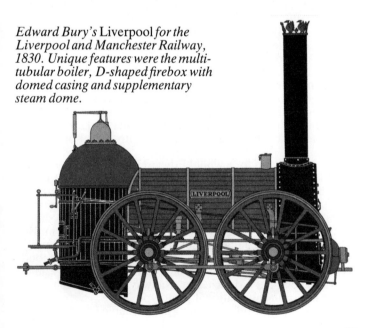

Edward Bury's Liverpool *for the Liverpool and Manchester Railway, 1830. Unique features were the multi-tubular boiler, D-shaped firebox with domed casing and supplementary steam dome.*

A modest beginning, perhaps; but in 1833 the line reached Hamburg, across the river from Augusta, Georgia. The distance of 135 miles made this the longest railroad yet.

Someone called the "Best Friend" the "Rocket of America", with adequate reason. But like the *Rocket* in England (which was later in two bad derailments) she was accident-prone. Her fireman was a Negro boy not sufficiently aware of the power of steam. When the safety-valve annoyed him with its din, he tied it down on one unlucky day, which happened to be June 17, 1831. The little hock-bottle boiler exploded. Darrell, who was in charge and might have seen his fireman's fatal prank, was badly scalded. The fireman died.

The "Best Friend" resembled none of the engines at the Rainhill Trials in England, save that like *Sans Pareil* she had her four wheels coupled. The cylinders were inclined, and between the frames, driving a crank-axle at the rear. These, and a water-tank below, balanced the vertical boiler, or were supposed to do so. After the accident, the engine was rebuilt and appropriately renamed *Phoenix*. In this form, the boiler was between the axles. It was still bottle-shaped, but this time suggestive of gin rather than German wine.

By then, other engines had come from West Point Foundry, mechanically similar but different as to their boilers. The second, the *West Point*, had a more-or-less English-style horizontal boiler, which was eventually to become the standard form in spite of American liking for the vertical type.

Here let it be said, with no intended insult, that early American locomotives were often very roughly made compared with those in England; but still they went. They were sturdy, indeed, and arranged as far as possible so that in the event of breakdown, repairs could be adequately carried out in a remote place by Basil the Blacksmith (cf. Longfellow, again!).

Back to England for a while! Edward Bury, as we have seen, was too late to produce a satisfactory locomotive for the Rainhill Trials, and just what his first engine, the *Liverpool*, was really like, we cannot show here, not knowing for certain. Probably she was not entered for the competition simply because she would not go in her original, mysterious form. But in partnership with James Kennedy (he who, most unhappily, ran down and killed Huskisson at Parkside in 1830) he rebuilt his engine in the form shown. The locomotive certainly ran on the Liverpool and Manchester Railway but is much more important as an international prototype.

Passing over her relatively enormous coupled wheels (quite

Robert Stephenson, 1803 – 1859, was the only son of George Stephenson, who made sure that Robert received the education which he missed. In 1821, he was assisting his father in surveying the Stockton and Darlington Railway and from then on he became more and more his partner rather than assistant.

freakish for 1830, though a great precedent) we see most prominently her inside bar frames or iron as well as her inclined cylinders below the smokebox, also inside. We see a horizontal boiler with a D-shaped firebox below a domed outer casing. Compared with either the *Rocket* or the *Best Friend of Charleston,* she is almost familiar! (A note in passing; the ornate red crown on the top of her stack was a cut-out procession of *liver birds*—mythical winged creatures who formed the insignia of that great and terrible city of Liverpool. The *liver,* in this connection, had nothing to do with an important organ of all the vertebrata, including ourselves!)

The *Liverpool* is extremely important in the development of locomotive design. Those bare frames were to be universal in America for as long as the steam locomotive lasted, while Continental and Eastern Europe, with much of the rest of the world, were to take them up sooner or later. The D-shaped, dome-topped firebox also was taken up by America, and was still to be seen in the eighteen-sixties, though not much later. All this arose out of some engines which Bury made for the Philadelphia and Reading Railroad (one of them was named *Rocket!*) in the eighteen-thirties. But the development of the Bury locomotive can keep a-while, while we look back at the Stephensons.

We left them building "Rockets", making some improvements within the limits of that very limited design. But, still in 1830, they produced the *Planet.* Her boiler was something like those of *Northumbria* and *Majestic*, the last of the Rocket class on the Liverpool and Manchester Railway. But the cylinders were inside and horizontal. The frames were outside, making a very strong and sturdy engine.

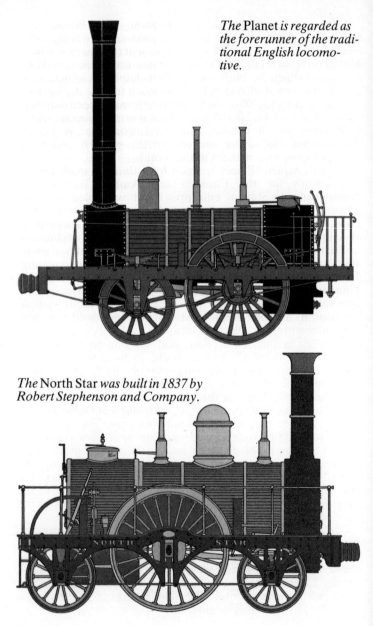

The Planet *is regarded as the forerunner of the traditional English locomotive.*

The North Star *was built in 1837 by Robert Stephenson and Company.*

Very quickly, improvements were made in this Planet type, the first of which was the use of four-coupled wheels for a freight variety. Various "Planets" were sent to America, where Matthias Baldwin copied it in a rather famous engine called *Old Ironsides*, which he built for the Philadelphia, Germantown and Norristown Railroad in 1832. So much trouble did he have in at last getting $3,500 of the $4,000 previously agreed that he is reported to have said in despair: "That is our last locomotive." It was not so, of course; Baldwin Locomotive Works of Philadelphia was to become one of the greatest and most famous locomotive makers in the world.

A four-coupled "Planet" named *John Bull* was made by the Stephensons for the Camden and Amboy Railroad in Pennsylvania (a Stevens enterprise) and assembled by Isaac Dripps in 1831. The engine had a dome-cased firebox in the Edward Bury style. Later, Dripps furnished a pilot, running on an axle of its own in front, which both steadied the engine and saved it, one supposes, from derailment if it encountered some absentminded cow; hence the widely used, if unacademic term *cowcatcher* for a locomotive pilot, current over about a century. Though the pilot was to become universal in America, it was ever rare in Europe except in the East and North. *John Bull* survived to become the oldest original locomotive in the Americas and is treasured to this day.

The South Carolina, *built in 1832, was the world's first articulated locomotive.*

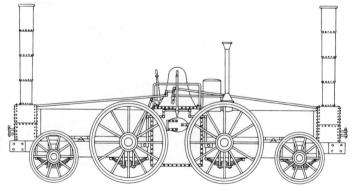

The first public railway to be opened in the German States was the Nuremburg-Fürth Railway on December 7, 1835. Our drawing shows the Adler, *the inaugural engine.*

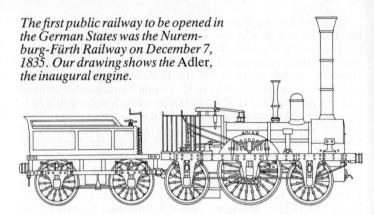

THE PATENTEE TYPE

"Planets" were unsteady at any speed, and their firebox capacity was limited. Stephensons rectified these by adding a trailing axle behind the firebox, which then could be made considerably bigger, in the early eighteen-thirties. From the first engine of the type, it was known as the "Patentee" class. The extra axle could be a free one on a passenger engine, or

The Monster *was built for the Camden and Amboy Railroad, c. 1834.*

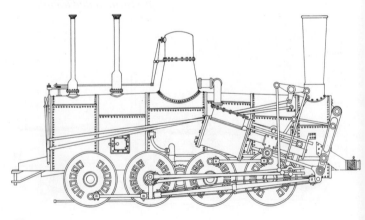

coupled for goods. Alternatively, the leading axle, or all three, could be coupled. "Patentees" of one sort and another were freely built for European service, as, one by one, the Continental countries essayed the new form of conveyance. The first locomotives for public service in Belgium (*La Flèche*, *Stephenson* and *L'Eléphant*, Brussels-Mechlin State Railway, 1835), in the German States (*Adler*, Nuremberg-Fûrth Railway, 1835), in Russia (Pavlovsk-Tsarskoye Selo, 1837), the Netherlands (*De Snelheid* and *De Arend*, Holland Iron Railway, 1839) and the Italian States (*Bayard*, Naples-Portici Railway, 1839) were all "Patentees". In its native England, the type abounded, and direct derivatives, still with wood-and-iron sandwich frames, were to be found there even in the early years of the next century.

An extremely fine example is the *North Star*, built by Robert Stephenson for the Great Western Railway in England in 1837. This splendid line was engineered by Isambard Kingdom Brunel, son of an emigré French father and an English mother. Brunel conceived railroads on the grand scale. By this time, railway construction was having its first boom and lines were being promoted and built in many parts of the Western World. The commonest rail-gauge was 4 ft. 8½ in., or 1.435 metres, which was not only that of the old colliery lines in North-eastern England, but was even approximate to the

Two Daumier prints show how signals were given in the early days on the railways in France.

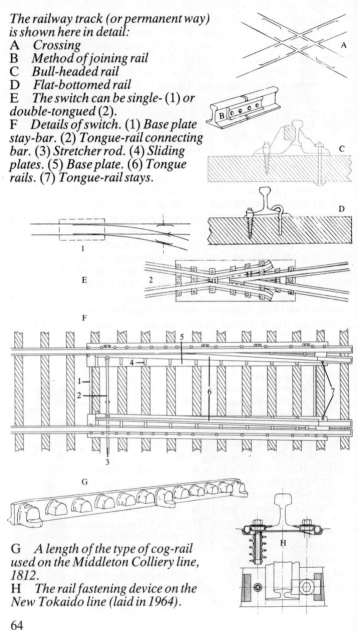

The railway track (or permanent way) is shown here in detail:

A *Crossing*
B *Method of joining rail*
C *Bull-headed rail*
D *Flat-bottomed rail*
E *The switch can be single- (1) or double-tongued (2).*
F *Details of switch. (1) Base plate stay-bar. (2) Tongue-rail connecting bar. (3) Stretcher rod. (4) Sliding plates. (5) Base plate. (6) Tongue rails. (7) Tongue-rail stays.*

G *A length of the type of cog-rail used on the Middleton Colliery line, 1812.*
H *The rail fastening device on the New Tokaido line (laid in 1964).*

The immortal Lion *was built over 130 years ago and is still in working order.*

wheel-width of the Roman roads in the days of the Caesars. Brunel alone thought that very inadequate to such an advanced conveyance as the train, and fixed his gauge at seven English feet, 7 ft. 0¼ in. to be exact, or 2.14 m. It is one of the tragedies of world railway construction that this did not become a generally used gauge, bearing magnificently spacious trains, though to be sure it would have been unwieldy in the narrow places of mountain country. As it was, it remained confined to the Western portions of Great Britain, and the last of it vanished in 1892. Stephenson's old gauge became widespread in Europe from the Russian borders to the Pyrenees, and ultimately throughout North America. We shall meet with wider gauges, though not on Brunel's grand scale, and many narrower ones. The Holland Iron Railway began with two metres (measured from centre of rail-head). Important broad-gauge countries today include the U.S.S.R., India, several South American republics, and Spain and Portugal.

Quite apart from its broad gauge, *North Star* was a large example of a "Patentee" (as *Adler* in Germany was a small one). Indeed, she was a big engine by any standards of the late eighteen thirties. Originally built for the 5 ft. 6 in. gauge New Orleans Railway in the U.S.A., a broken contract was the cause of her being altered to seven-foot gauge and going to the Great Western instead. She had driving wheels 7 ft. in diameter and the inside cylinders were 16 in., both diameter and stroke; total heating surface was about 711 sq. ft. Obsolescence was rapid in those days, but the *North Star* was rebuilt with new cylinders and a larger boiler in 1854, and was in service for over thirty-three years, a prodigy of that period. She had taken the first Great Western train out of London, on June 4, 1838. She was preserved by the Great Western company for many years. A barbarian mixture of indifference and vandalism brought her to the scrap-heap in 1906, since when many specious excuses have been made, and a partly wooden replica constructed.

Much happier in such respect was the story of the *Lion*, a front-coupled "Patentee" with a square Gothic-arched firebox casing, built as their first engine by Kitsons of Leeds (Todd, Kitson and Laird) during the winter of 1837-38 and bought by the Liverpool and Manchester Railway for freight traffic. The L. and M.R. was amalgamated into the Grand Junction Railway which had come to connect it with Birmingham, and these in turn, by amalgamation with the London and Birmingham Railway, the first great line out of the capital, became parts of the mighty London and North Western Railway, which in turn inherited the *Lion*. The L.N.W.R. sold her for docks service in Liverpool, where she latterly (indeed for many years) worked a dock pump. From this she was rescued in the nineteen twenties and was fully restored by Robert Stephenson and Company in 1929, under the direction of A. C. W. Lowe. She steamed again at the Liverpool and Manchester Railway Centenary celebrations in 1930. Perfectly sound, she has appeared in various films since then.

As remarked, the wheel arrangement of a "Patentee" was varied in Europe from 2-2-2 (1-A-1) for passenger haulage to 0-4-2 or 2-4-0 (B-1 or 1-B) for heavier work requiring more adhesion. Also remarked previously was the early incidence of its predecessor, the "Planet" type, in North America. It was here that certain very interesting, non-European, variations took place in the basic "Planet/Patentee" form. While wrought-iron rails, at first chaired on stone blocks and then on timber sleepers or cross-ties, were general in Europe, America clung doggedly for a long time, on certain lines, to the wooden road with iron straps on top of the baulks; also, these early American roads were often laid very roughly on a road-bed scarcely worth the name. At the same time be it added that proper English iron rails were extremely expensive. In an expanding country, the thing was to get the railroads down as quickly as possible so that new towns and plantations could blossom beside them. That roughness demanded a much more flexible locomotive if the daily train was not to spend much of its time off the rails and possibly axle-deep in good earth.

One way to this end was the use of a flexible leading truck on two axles, not so much as a pivoting agent on curves as to give three-point suspension to four wheels. Its origin has been argued. The invention has been claimed for the Stephensons (the English word "bogie" is a North-Eastern one for a truck). Certainly the appliance was first made in the United States,

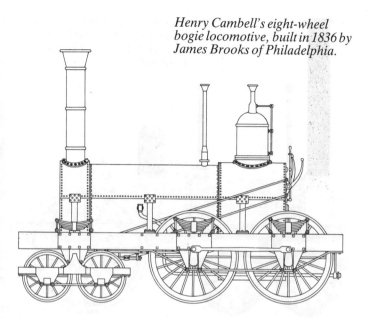

Henry Cambell's eight-wheel bogie locomotive, built in 1836 by James Brooks of Philadelphia.

wherever it was invented. In fairness be it remarked that a deputation of American engineers visited the Stephensons as early as 1828, and that Robert Stephenson spoke to them about the arrangement, which he had not yet built.

In 1832, John B. Jervis designed, and had built by West Point Foundry for the Mohawk and Hudson Railroad, an engine which he called *Experiment,* but which has gone down to history as *Brother Jonathan* (at that time the rather quarrelsome Anglo-Saxon cousins were known respectively as John Bull and Brother Jonathan; Uncle Sam and Wilhelm Busch's Mister Beef had not yet arrived properly).

This engine's likeness shows her arrangement sufficiently well. Basically, she was a "Planet" but she had the leading truck replacing the old rigid axle behind the smokebox, while the driving axle, instead of being close behind it, was in rear of the firebox, anticipating patents later granted in England to T. R. Crampton, of whom we shall see more later. She was a *very* small engine, but—probably with little load—she was regarded as a flyer. Few of the early speed records can be regarded as authentic, and *Brother Jonathan*'s (or *Experiment*'s) mile-a-minute must stand in the apocrypha. It *might* have been so.

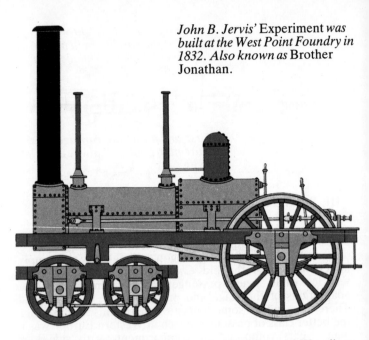

England built a similar engine for the Camden and Woodbury Railroad in the United States in the following year of 1833, when, also, J. and C. Carmichael in Scotland built a freakish one (0-2-4 with side-lever drive) for the Dundee and Newtyle Railway.

A much more interesting American development of the basic Stephenson type was that of Henry Campbell in 1836. For the Philadelphia, Germantown and Norristown Railroad in Pennsylvania he built a locomotive which had the general arrangement of a "Planet" but with the Stephenson-Jervis truck in front, and an extra axle in rear of the firebox (also Jervis) which was coupled to the driving axle by outside cranks and rods. The engine was not very successful, by some trustworthy accounts, and for years was somewhat forgotten. Certainly she did not become a great American prototype. But she was the first of the widely used 4-4-0 type as to wheel arrangement, and she had such mighty European descendants as the French "Outrance" bogie class in the 'seventies and even the very advanced English "Cities" of the nineteen-hundreds (Northern, and Great Western Railways respectively). In these, of course, there was no timber in the frames.

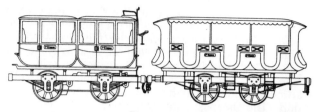

First- and third-class carriages on the Nuremburg-Fürth Railway in the late 1830s.

THE PASSENGER CAR

Before we go on to the development of the other early classic types such as the Bury with its bar-frames, let us look at car design. At the very first, the wagon or freight-car was simply a larger version of the old *Chaldrons*, the primeval mining vehicles, while ancient passenger coaches were flanged-wheel adaptations of the stage and private coaches on the roads. There were indeed some fantastic experiments, but in general the better class of passenger coach in Western Europe was for many years composed of compartments with side-doors: stage-coach bodies combined, three, four, or more at a time, on a single railway frame. This type made its appearance with the opening of the Liverpool and Manchester Railway, and probably originated with Nathaniel Worsdell, the eminent coachbuilder who had made the *Rocket's* tender. Each compartment, or "body", held six passengers on heavily

One of the carriages drawn by the Atlantic *in 1832 for the Baltimore and Ohio Railroad.*

padded but very upright seats (we have ridden in a replica!). Cheaper fares were charged for riding in an open-sided *char-à-banc*. Both types are shown in the drawing of the Nuremberg-Fûrth train, as well as in the famous Ackermann "Long Prints" of the Liverpool and Manchester Railway.

On the latter, for an extra charge, people might ride in compartments belonging to the mail carriage, providing corner seats only. The cheapest passengers rode in box-like wagons which were often without seats and known in England as "Stanhopes" (a very bad pun on "Stand-ups"). The Stanhopes were an old and honoured family, so presumably their name was borrowed to lend these frightful vehicles a spurious sort of respectability. One doubts that it had anything to do with the Stanhope and Tyne Railway, which had no monopoly of this mean sort of "carriage".

The side-door-compartment type of coach, however, answered well for many years. Before very long it was being used—with proportionately narrower compartments and harder seats—for second- and third-class passengers. A century later, it was still being built for suburban trains around the big cities of Great Britain on account of its high capacity and its ease of entry and exit, though of course it was by then a very much larger thing than its remote precursors had been.

A favourite arrangement among the richer English families of those early days was to ride in their own private coaches chained to flat wagons. It was "exclusive"; ladies and gentlemen needed not to soil their skirts and pantaloons by sitting on public cushions; but the dust, fumes and cinders were the devil to pay when the conveyance was an open landau in summer. The Duke of Wellington was recorded as having travelled thus when he was over eighty, though the South Eastern Railway had thoughtfully made a special railway carriage for him. Probably the last person to travel so was a most unpleasant English eccentric, Mrs. Caroline Prodgers, who was observed *en route*, covered with dust, at Chesterfield in the English Midlands as late as the eighteen-eighties, but there is a sort of revival in our own time, where motorists and their cars are ferried by train.

The cheaper passengers had an awful time in the "Stanhopes" and other abominations. Throughout Europe it was much the same. An old French cartoon (by Daumier) shows the third-class passengers being lifted out by porters, stiff as frozen codfish.

Though all European countries at that time regarded the American way of life as rough and horrid, to America railroad

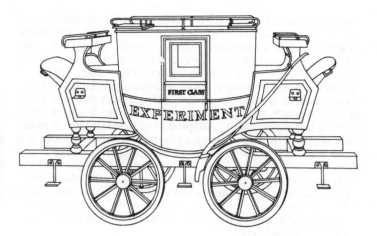

The Experiment *carriage was built for the Stockton and Darlington Railway in 1825.*

Early stage-coach type American car.

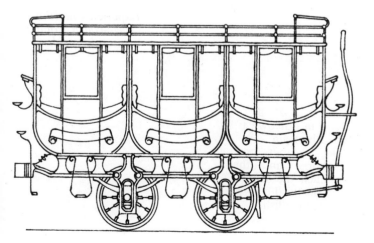

history are owed some considerable debts for introducing, at one time and another, passenger cars that were at one more practical, and occasionally more comfortable, for long journeys. Once again the stage-coach furnished an early model, but even here there were differences. The American stage-coach was a longer, more substantial vehicle than the sporty British article, though less refined than the sometimes massive French *diligence*. It had to negotiate frightful roads without too many upsets. Adapted to railroad service, it might have seats on top, as in Imlay's coaches for the Baltimore and Ohio Railroad, or have an almost boat-shaped body, as in our drawing of *Old Ironsides* with her train. Both these and the early British railway carriages suffered from their short wheelbase, which gave them a nasty fore-and-aft motion, called with some reason by children, "sick-making". As might have been expected, one of the first improvements was in mounting each on eight wheels.

A very early example of a railroad coach on pivoted trucks or bogies was one called *Victory* built by Imlay for the Philadelphia and Columbia Railroad in 1834. (Before this, there had been one on the St. Etienne-Lyon Railway in France.) An old model of *Victory* survived the years. Between the trucks, the body was of Imlay's boat-shaped form. An apparent clerestory deck on the roof may have been a sort of box-truss to prevent sagging. There has been speculation about apparently blind compartments over the trucks, even so

Experience *had three "stage-coach" compartments side by side.*

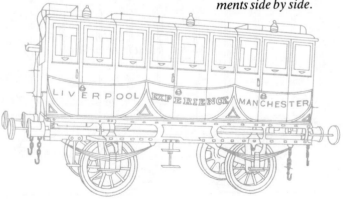

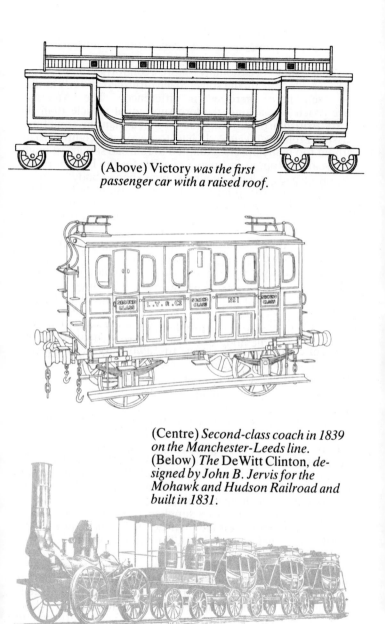

(Above) Victory *was the first passenger car with a raised roof.*

(Centre) *Second-class coach in 1839 on the Manchester-Leeds line.*
(Below) *The* DeWitt Clinton, *designed by John B. Jervis for the Mohawk and Hudson Railroad and built in 1831.*

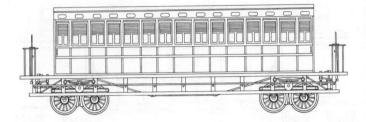

*American passenger car
designed and built by Ross
Winans, early 1840s.*

optimistic as to suggest that one contained a bar and the other a
water-closet. The likelihood seems remote. But eight-wheelers
appeared generally in the 'forties under the influence of John
Stevens on the Camden and Amboy Railroad, and Ross
Winans on the Baltimore and Ohio. By the time Charles
Dickens came to the United States in 1841, he was already able
to describe their coaches as resembling a *shabby omnibus*, with
a passage down the middle. His hosts, who had welcomed him
as a distinguished English Liberal, were annoyed when they
read his *American Notes*.

By then there were even sleeping cars of a sort. On the
Cumberland Valley Railroad in the United States they put on
bunk-cars which enabled people to lie down by night, though
not, possibly, to sleep. Even at the speeds of the day, the
journey was short. The time was 1836.

Two years later, in England, there was the *bed-carriage* on
the night service from London to the North West. It was of the
usual compartment type with, at one end, a boot such as the
road coaches had had for carrying the mails. But in this case, the
end passenger partition was hinged; padded boards were put
between the seats, and for a supplement first-class passengers
could lie down with their feet in the extra space. If they did not
literally go to bed in their boots, two or three might do so with
their feet in a collective one.

"First-class!" The term arrived early, as did "second-class"
and "third-class" where Europe was concerned, to denote
excellence or inferiority according to the fare paid. But in
America then, as in communist countries much later, "class"
was a dirty word. So old American railroad advertisements
advised patrons that the fare would be so-many dollars in the

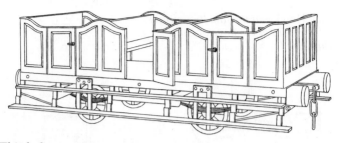

Third-class carriages were introduced in Britain in 1838.

best cars and so-many-less in the *accommodation cars*. Even under Louis-Philippe, that uneasy French king, egalitarian feelings were soothed by the terms *diligence*, *omnibus* and *wagon*. That must do for passenger cars for the present. Except in America, they were quite singularly unimaginative; on the other hand the European first-class soon became very comfortable within the limitations of tradition, especially on the broad-gauge trains of the Great Western Railway in England, whose standard first-class compartments could seat eight assorted stout men and fat women—John and Joan Bulls—with room to spare on the best hairstuffed morocco leather. But there was no winter heating for them, no applied sanitation, while the vegetable-oil lamps dropped through holes in the roof at night were of the dimmest sort. These appeared in the late eighteen-thirties.

ENGINEERING PROGRESS

From these passenger-miseries, let us turn to the more heroic progress of locomotion, and at this point we come to some singular equations of English, American and German practice. In England, the firm of Bury, Curtis and Kennedy in the late 'thirties was building Bury's little bar-framed locomotives, for the London and Birmingham, and other English railways, and for the American market. In America, indeed, they did quite well, though the Baltimore and Ohio Railroad was sticking to the old vertical type engines with bottle boilers, exemplified by the "Grasshoppers" of Phineas Davis (so-called because their overhead levers suggested the hind-legs of those agile insects).

75

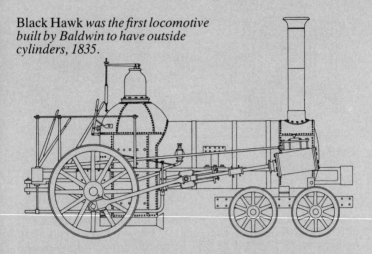

Black Hawk *was the first locomotive built by Baldwin to have outside cylinders, 1835.*

But the Bury engine was still a rigid thing for American tracks of the time. Matthias Baldwin's eleventh locomotive, the *Black Hawk* of 1835, had a Bury boiler combined with inclined outside cylinders at the front, which was supported on a four-wheel truck, while the driving axle was in rear of the firebox, as in *Brother Jonathan*. There were no main frames; the boiler barrel held the engine together, as with the steam road-rollers of much later years.

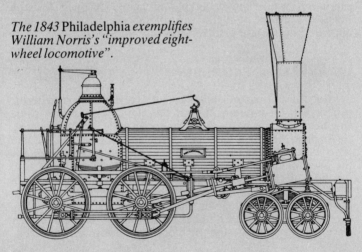

The 1843 Philadelphia *exemplifies William Norris's "improved eight-wheel locomotive".*

Another Norris engine, the George Washington, *built in 1836.*

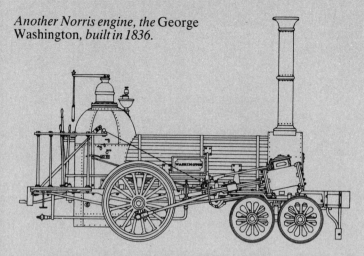

Even more successfully, William Norris of Philadelphia took the Bury locomotive, retaining the bar frames and the driving axle in front of the firebox, but using the leading four-wheel truck and inclined outside cylinders as in the Baldwin *Black Hawk*. Norris's *The Washington Farmer* of 1836 distinguished herself by hauling a load up a gradient of 1 in 14. The secret of her success was doubtless in the high proportion of her modest weight being on the driving wheels. The Norris locomotive

Lafayette, *built in 1837 in Philadelphia, was the Baltimore & Ohio Railroad's first engine with a horizontal boiler and more than four wheels.*

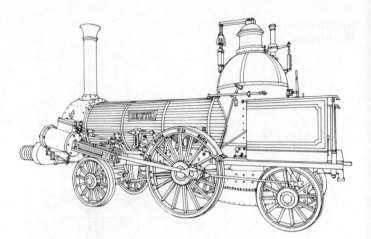

The Beuth *was built in 1844 by Borsig for the Berlin-Anhalt Railway.*

of this type was extremely successful in its day; even the Birmingham and Gloucester Railway in England invested in a set and had more built in British works.

Development became rapid. In the *Hercules*, built for the Beaver Meadow Railroad in 1836, two coupled axles, fore-and-aft of the firebox, replaced the single one. Garrett and Eastwick of Philadelphia were the builders; their foreman was Joseph Harrison, who had been one of Norris's men. He became a partner, and in 1839, Eastwick and Harrison produced the *Gowan and Marx* for the Philadelphia and Reading Railroad. This was of pure Bury-Norris type, but with both coupled axles ahead of the firebox. There was a steam-jet blower for maintaining draught while standing. Hackworth had first used this, but it was hitherto unknown in the States. Though she weighed but eleven tons, the little *Gowan and Marx* succeeded in hauling 423 tons on a slightly falling grade from Reading to Philadelphia at nearly ten miles an hour. It was a prodigy!

Norris's famous design had developed, by 1839, into the form exemplified by his No. 25, named *Pegasus*, for the Baltimore and Ohio Railroad. This was to be the basic form of the ordinary American locomotive for half a century, though

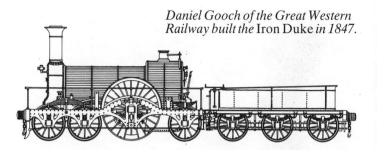

Daniel Gooch of the Great Western Railway built the Iron Duke *in 1847.*

progressively enlarged and improved. Indeed, it became cosmopolitan.

A most interesting development of the Bury-Norris form belongs to Germany. August Borsig founded locomotive works first at Moabit and then at Tegel, Berlin, and then built his own foundry at Moabit. He took Norris's six-wheel type, but to steady it he added a trailing carrying axle. Higher speeds were intended, without coupling the axles. This, and many succeeding engines, were built for German service initially for the Berlin-Anhalt Railway and later for the Cologne-Minden Railway and others. Now German railways at the time were closely modelled on the British, with a fairly solid *permanent way* compared with the happy-go-lucky American *track*. While that leading truck had been designed to cope with the inequalities of the latter, it showed on the former, at higher speeds, a fearsome tendency to waggle about. At worst it could completely derail and get itself at right-angles to the engine frames. In later designs, Borsig substituted a single rigid axle in rear of the cylinders, and on the solid road all was well again. This form may be studied in a splendid replica of Borsig's *Beuth* (named affectionately after his old professor) in the Deutsches Museum, Munich. *Beuth* was altogether a more finished, more robust and indeed very beautiful engine compared with the original *Borsig*.

A few notes are deserved on early German railways generally. Germany was still a sprawl of independent states ranging from kingdoms to tiny principalities, not yet welded into an empire by Prince Bismarck and the King of Prussia. They ordered things very much their own way, with dozens of separate railway undertakings under company ownership. Soon after Queen Victoria had succeeded to the British throne and married Prince Albert of Saxe-Coburg-Gotha, there was a

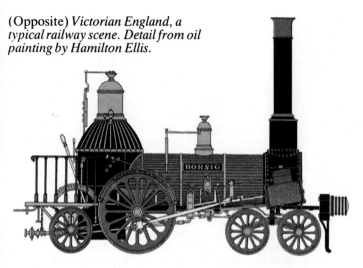

August Borsig's first locomotive, 1841, took the American Norris type as its model.

Förstlingen *was the first locomotive to be built in Sweden, 1848.*

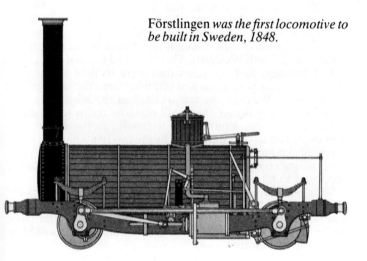

81

family squabble with Ernest-Augustus, King of Hanover and Duke of Cumberland. Under the Salic Law, Victoria could not succeed to the Hanoverian throne (as the Queen delicately put it: "Hanover is no longer a British possession!"). But Ernest-Augustus equally could not succeed to the British throne, and one retains an impression of a king with a chip on his shoulder. According to a persistent story, within his kingdom he decreed that trains entering Hanover should *never connect* with trains going to a further destination. There was business, however, in his cantankerosity. It meant that travellers from West to East, kicking their heels in his capital, were bound to spend money there.

In a previously distrustful world, railways enjoyed in the eighteen-thirties a business boom. In the middle eighteen-forties there was a second, far greater one, known in England with some reason as the Railway Mania, which was reflected on a rather less lunatic scale in other countries. Sanity prevailed in Belgium, already quite highly industrial and densely populated, where railway construction from the first was under State enterprise and planned for the national advantage. Whether the promoters were acquisitive or State-minded, it resulted in important railways being built between principal cities and out to other places. The steam railway became a part of modern life.

Under such conditions, it flourished, and ironically it advanced most where people believed in cut-throat competition, the Law of the Jungle, and all that.

As to the train, let us turn back to that magnificent Great Western Railway in England (which was to be speedily at war with its neighbours, the London and North Western and the London and South Western). The three had the first main lines out of London, and for years they were to fight bitterly for traffic in overlapping areas and often between the same cities.

The Great Western, as remarked, had the advantage of a most generously broad gauge, allowing its trains to be much faster and very much more comfortable for its passengers. On the other hand it was sandwiched by the two rivals named. Even in practical America, there were at first variations in gauge, and a competing company would virtuously advertise itself as being "opposed to all monopolies".

So to the Great Western Railway in England let us go for what might be called the *Apotheosis of the Patentee*, though it was not quite that. It was the transformation of a little basic design (e.g. Bavaria's *Adler*) into what for many years would be a *very* big locomotive.

Daniel Gooch, a North-eastern Englishman, was still under 21 when I. K. Brunel placed him in charge of the Great Western locomotives, which apart from *North Star* and *Morning Star* were a frightful outfit of mechanical monsters. The young Gooch wisely standardized on the Stephenson Patentee type with its solidity and what one might call its mechanical integrity. His brother John assumed a similar position on the rival South Western line. As far as we know, the brothers, though under business rivals, between them produced a valve gear which, through variation in the position of a link between excentrics and valve rods, could use steam expansively instead of simply admitting and exhausting it, as in the old "gab" gear which survives to this day in Kitson's *Lion*. Almost simultaneously, Robert Stephenson and Company, and Alexander Allan on the Grand Junction Railway, also in England, were working on similar gears, all of which were to be used for many years all over the world.

As to straight locomotive design, by the late eighteen forties Gooch had produced the first of many splendid eight-wheel engines—still basically "Patentees", but incomparably bigger, more powerful, and faster. We show the *Iron Duke* of 1847 traced from a plate in the 1851 edition of Tredgold's *The Steam Engine*. This engine had 8 ft. driving wheels, 18 in. by 24 in. cylinders, 100 lb./sq. in. boiler pressure (later 115 lb.) and 1,944.8 sq. ft. of heating surface, so she was indeed a giant of her day. Corresponding vital statistics of the original *Patentee* were 5 ft., 12 in. by 18 in. and 50 lb./sq. in.

The *Iron Duke* was named for Wellington, that eminent general of whom it was written that *the more one reads of Wellington, the more one respects him and the less one likes him*. He was a military hero in the same company as Leonidas, Pompey, William Wallace and Gustavus Adolphus, and it was not through his own fault that, unlike them, he died in his bed. As to the engine, many like her were built, with but minor alteration, for forty-one years after.

One of the class, the *Great Britain*, made in May, 1848, one of the earliest record runs, fully authenticated by instruments on a dynamometer car (also Gooch's invention). She took a train of ordinary coaches plus the dynamometer-car from the original Paddington terminus in London to Didcot, 53 miles, in 47.5 minutes. The train probably weighed about 75 tons at most, but the average speed, start-to-stop, was just over 67 miles an hour, against a slight but constant gradient. Ordinary Great Western fast trains at the time averaged rather more than 50 miles an hour, start-to-stop, with usual maxima in the 60s. These speeds

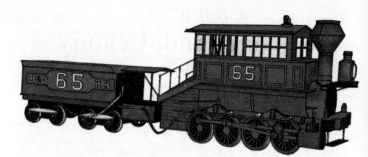

*The Baltimore & Ohio
Railroad's 0-8-0 no. 65,
built in 1848, is an early
example of the American
"Camelback" type.*

were much faster than anything else in the British Isles; indeed the Great Western, with its great engines on broad gauge and an extremely solid road, was by far the fastest railway in the world at that time. It extended from London to Bristol, with broad-gauge connection by allied companies to Exeter and even to the outskirts of Plymouth at Laira. It was indeed a majestic railway.

*T. R. Crampton's famous design, built
in 1846.*

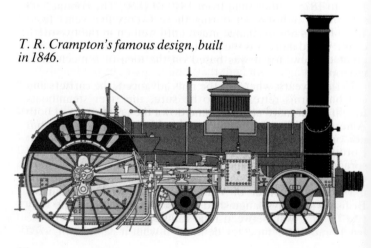

Chapter 4
Mid-Nineteenth Century

From the railroading point of view, this term means the years 1850 to 1875, rather than from 1840 to 1860. The commercial side of the industry set during those twenty-five years from 1850. It was not to change much until well on in the twentieth century, and more was the pity from the practical railwayman's point of view, for it was based on the idea of a mechanical monopoly.

In those years, wherever the rails advanced, the carriers and coaching firms retreated. To be sure, the river steamboats fought a rearguard action up and down the great rivers of both America and Europe. They could hold their own on bulk but not on speed, as indeed they continued to do on the great European waterways. As for transport by rail, the lines had spread through the land of their birth from south-western Cornwall (Trevithick's country) to Caithness (the extreme north-eastern corner of Watt's Scotland) while the narrow British waterways languished and sometimes died. The only real fighting was between rival railway companies where the local brand of democracy demanded what it called *unfettered competition*, as in North America, and to a lesser extent in Great Britain. There were several large-scale brawls between company-hired ruffians in the latter; in America there was sometimes not only battle but murder and sudden death, with charging locomotives striking on disputed rights-of-way and guns being fired in anger.

Assisted by such various agents as American gunmen and that Emperor of All the Russias, who is supposed to have decided a disputed survey by using his sword as a ruler for the line from St. Petersburg to Moscow, the lines spread rapidly. They were all over Western Europe, even over the Semmering and under the Col de Frejus. They spanned European Russia, even in the deep south. They were opening up Scandinavia where, ever since 1798, they had been linking waterways and mines, though steam did not appear until Munktell of Eskiltuna produced the grotesque little *Förstlingen* (*The Pioneer*) in 1853. Most spectacularly they crossed North America from Atlantic to Pacific (1869).

Such tremendous geographical development meant corresponding mechanical improvement and variations. Railways across mountains needed locomotives very different from those on fast service between cities; more akin to those for heavy coal and iron haulage. One cannot easily divide mechanical history by arbitrary dates, and here we must look back a little. From

Ross Winans' "Mud-digger" type.
That illustrated is the Cumberland.

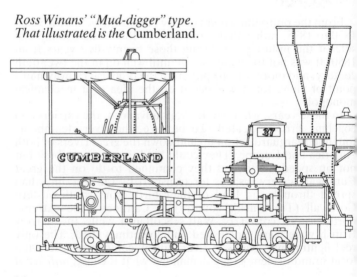

early Victorian England, with the astonishing flight of that auspiciously named engine *Great Britain*, we must turn first to America and then to Austria.

Something odd was not unexpected on the old Camden and Amboy Railroad in Pennsylvania with Isaac Dripps in action. His *Monster*, 1836, had been a monster indeed. The drawing is fairly explanatory, when one recovers from astonishment. Points to be noted are the drive through side-levers by reversed cylinders, and the effect of eight coupled axles achieved by coupling two pairs and putting gears between Nos. 2 and 3. Very notable was the use of tapering in the boiler shell, which was to be regarded as "modern" a century later, at any rate in the English-speaking countries, and for long in Austria too.

Again among eight-coupled coal engines, there were the "Mud-diggers" of the Baltimore and Ohio Railroad. The second one, with a Bury boiler, is shown here. She was built by Ross Winans in 1844, and her predecessor had a large version of the old upright gin-bottle boiler, which was hopelessly inadequate for any but a very small engine. Both had simple coupled axles, though drive was through jackshaft and sput wheels.

Both these types, with their rigid wheel-bases, were prone to derailment on the light tracks of the time. The nickname of the Winans engines describes their habits only too well, while Dripps' *Monster* was later rebuilt as 4-6-0, or 2-C. But long before that, the idea of a really flexible engine had engaged engineers. Even Blackett and Hedley had tried it, and in 1832, for the South Carolina Railroad, West Point Foundry built an experimental engine with twin chassis supporting a double-barrelled boiler, the firebox in the middle. The design is usually credited to Horatio Allen. It was a very early example of an articulated locomotive, assuming *Puffing Billy*'s metamorphosis to have furnished the first.

THE SEMMERING TRIALS

Our scene shifts to the old Habsburg Empire, where between Vienna and Trieste was built the first great main-line railway across a major mountain range. The Semmering Pass went back in transport history to the Middle Ages. It was the lowest of the Alpine passes, but still, for a railroad, involved a summit level of 2,880 ft., much tunnelling, and a ruling gradient of 1 in 40. The line was begun in 1848, by Carlo Ghega, who favoured locomotive traction from the first, though State bigwigs

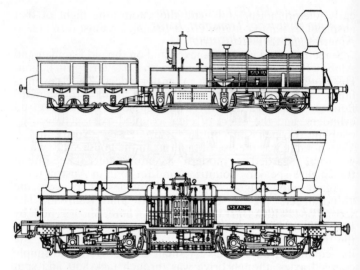

(Above) *Winner of the Semmering Trials, the* Bavaria.

(Centre) Seraing, *Belgium's contestant at the Semmering Trials.*

(Below) Vindobona, *built by John Haswell for the Semmering Trials.*

88

*Wilhelm Engerth's 1851 design for the
Austrian Ministry of Transport. It was
an excellent mountain locomotive.*

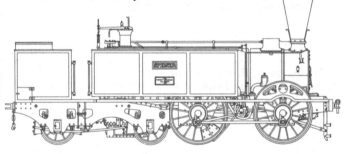

*Wilhelm Engerth's express tank
locomotive,* Speiser, *built in 1857 for
the Swiss Central Railway.*

suggested both atmospheric and cable traction as being the only
ones possible. Atmospheric traction, be it added, was a
mechanical aberration of the middle 'forties. It involved a
continuous vacuum pipe with a slot in the top, closed, or
supposed to be closed, by a flap-valve of greased leather. Trains
were connected to pistons running inside the pipe, and while
stationary pumps at intervals exhausted the air in front of the
train piston, atmospheric pressure in its rear pushed the cars
along at very considerable speed. The cumbersome locomotive
engine, said its promoters (among whom one is sorry to count I.
K. Brunel) was doomed. Four such lines were in public service:
London-Croydon, and South Devon, in England; Le Pecq-St.
Germain in France, and Dublin-Kingstown (Dun Laoghaire) in
Ireland. Apart from absurd junction complications, the thing
was a frightful fiasco after a very short time. Wear-and-tear was

fearsome. Rats ate the greased leather valves with great relish. It came and it went, most fortunately, before people could take it and play pranks on that magnificent Austrian Southern Railway over the Semmering.

A prize of 20,000 Florins was offered for a powerful and flexible locomotive capable of working such a line. The competition has been called, with retrospect to that between Liverpool and Manchester in 1829, the "Rainhill of the Alps". Joseph Anton Maffei of Munich, whose locomotive genius was Joseph Hall, an emigré Englishman, entered the *Bavaria*. She was the winning engine and managed 132 tons behind the tender at 11.34 miles an hour on the rising gradient of 1 in 40. But she was really a box of tricks. Her second and driving axles were chain-coupled as in ancient Stephenson practice, and so were her trailing axle and the leading tender axle. Side rods coupled the leading and second axles. The chains constantly broke; the engine did no regular work, and her big boiler was used to drive the machines in the workshops at Maribor (now in Yugoslavia) until 1870.

From John Cockerill of Seraing, Belgium, came the interesting little double engine *Seraing*. Here was a relatively modern development of Horatio Allen's *South Carolina*, with two motor-bogies and a centrally-fired, double-barrelled boiler. In its turn, this design anticipated the double-bogie engines built in quite considerable numbers by British works from the middle-sixties onwards under patents granted—with rather dubious justification—to Robert Fairlie. Fairlie used outside cylinders, which made his engines much more accessible than Cockerill's.

John Haswell was a Scot, born near Glasgow in 1812. He went to Austria about 1837 and soon became engineer of the Vienna-Raab Railway and Director of the Imperial-Royal Austrian State Railway Works. We have seen Haswell locomotives at work and in good shape when they were over a century old, on the Graz-Köflach Railway in Styria. That outrageous fraud of our own time, which some rascal called "built-in obsolescence"—meaning jerry-built to wear out quickly—was unthinkable in the great old days.

For the Semmering trials, Haswell built an engine called *Vindobona*. Our drawing shows her as she appeared during the trials; the first European locomotive with four coupled axles, and the forerunner of many thousands. One notes the firebox, which had a flat crown-plate and a rectangular casing, anticipating the form of Alfred Belpaire, that eminent Belgian. Though her Semmering performance was rather feeble, she was uite the most practical engine of those entered, with few roublesome gadgets.

Günther of Wiener-Neustadt entered another articulated ngine with two motor bogies like Cockerill's, but with a single ong boiler. It had outside cylinders, being thus better than the *eraing*, though the Stephenson link motion remained lreadfully inaccessible. It rather anticipated the Mallet and Meyer types of locomotive built in more recent times.

The Austrian Southern Railway bought all these engines, but ne doubts that any were of much more use than the winner, the ll-too-complex *Bavaria*, which showed the best trial erformance. All could do the job, provided that they did not reak down, but they often did so. Late in 1851, Wilhelm Ingerth made for the Austrian Ministry of Transport yet nother design. In this, the rear of the engine was partly upported by the tender through a radial connection ahead of he firebox, this coming between the two tender axles. The atter in turn were coupled by side rods, and a set of three gear vheels conveyed tractive power from the driving axle to the eading tender axle. This flexible drive to the tender was indeed he only weak point of the design. Though ingenious, it gave rouble like the other people's complex drives. Without it, the Ingerth locomotive made an excellent mountain engine of its lay, and in France the Eastern, Northern, Midi and Dauphiné Railways found it useful also for heavy coal haulage. There vere even Engerth express passenger locomotives in France ind Switzerland, and from the latter we show one of a once numerous class on the Swiss Central Railway. The *Speiser* was puilt in 1857 and ran until 1902, afterwards being kept as a relic. We recall seeing a six-coupled Swiss "Engerth" still in service as ate as 1927, and a French one worked on the La Bastide-Mende line up to World War II. The Engerth principle vas applied to narrow-gauge locomotives for Spain right into elatively modern times, and in a modified form. As we write, ve know of quite advanced examples still at work in that country.

T. R. Crampton's giant, Liverpool, *1848*.

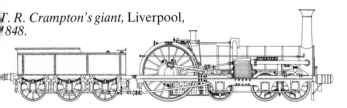

Austrian practice, however, discarded the supporting-tender principle fairly early, finding that a long-boilered eight-coupled locomotive (0-8-0, or D) served perfectly well the purposes of the Semmering, and later of the Arlberg, Brenner and Tauern lines.

Several funny things happened in British locomotive engineering at mid-century. We have seen Joseph Hall, the Englishman, going to Munich and producing *Bavaria*, we have seen Haswell the Scot becoming an eminent Austrian. Next we have the case of Thomas Russell Crampton, who was another Englishman, being cold-shouldered in his own country but honoured and rewarded by both the Germans and the French.

He had been one of Daniel Gooch's young men on the Great Western Railway, and his experience with broad-gauge engines convinced him that, subject to some mechanical tricks of his, an ordinary "narrow" locomotive could go just as fast, and even be as powerful, as Gooch's splendid examples. Some of his earlier patents and designs look fantastic, but by the late eighteen-forties he had produced a most interesting type of locomotive.

As will be seen, he placed his driving axle in rear of the firebox, to keep the centre of gravity down. (Low centre-of-gravity was a shibboleth of the period; it was supposed to prevent trains from tipping over on curves, but in fact it caused much rough riding and disturbance of the still flimsy tracks.)

*Two railway policemen
slowing down trains with
flag signals (1840s).*

THE CRAMPTON

Crampton's first practical locomotives were six built during 1847. Two, named *Liége* and *Namur*, went to Belgium. One, named *Kinnaird*, went to the Dundee and Perth Junction Railway in Scotland. The others became Nos. 81, 83 and 85 of the South Eastern Railway in England, where Charles Dickens may have been familiar with them, he knowing the South Eastern well. (It killed Mr. Carker, very properly, in *Dombey and Son*.)

Apart from its huge driving wheels right at the rear, the design had other features very important for the future. Cylinders, connecting rods and valve gear were all completely outside with the platforms raised to make them entirely accessible. Years after, this was to be invariable practice in both Continental Europe and North America, while ironically England was very late in adopting it.

Two rather similar engines to these were built; one generally larger and called *London* for the London and North Eastern Railway, and the last with a longer boiler barrel, for the Maryport and Carlisle Railway in the extreme North of England. In all cases the tenders seem to have closely resembled the standard form on the Great Western Railway, recalling T. R. Crampton's office under Daniel Gooch. These early Crampton engines were not entirely successful. Their boilers were inadequate, especially as to firebox heating

This 4-2-0 was built by Cail of Lille in 1859 for the Chemin de Fer du Nord. Designed by T. R. Crampton.

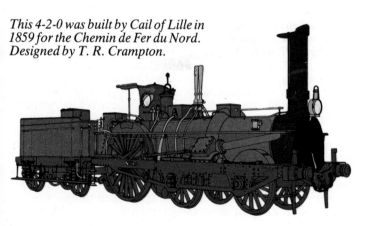

surface: indeed the original Crampton firebox was an odd thing, with the grate longer than the distance between back-plate and tube-plate. The Scottish engine *Kinnaird*, however, lasted long enough to be photographed in service by R. E. Bleasdale, the first photographer to make a systematic collection, recording locomotives all over the British Isles during the second half of last century.

Crampton's patents were many, and he was responsible for several distinct types of locomotive, some with drive through jackshaft or "dummy crank-axle".

By far the most successful was that generally called the "French Crampton", which had very strong outside frames, the cylinders being supported between these and the inside frames, together with the driving and valve motion. *Le Continent* of the Eastern Railway of France was built in 1852 by Derosne, Cail and Company of Paris, who had first taken up the design in 1848 with an order of twelve engines for the French Northern Railway. With a light express train of first-class carriages, forerunner of the "limited" of later years, a "French Crampton" was an extremely fast runner. Though rather rough-riding, with their low centre of gravity and relatively long rigid wheelbase, the engines were steady runners, and were much liked both in France and in the German States. That named above was in service until 1919 on light work, any locomotive being useful in wartime! The old thing had been in the retreat after Sedan, back in 1870. She can still steam, and occurs in films from time to time.

Ironically, though even the "French Crampton" was initiated with several engines on the English Eastern Counties Railway, and one on the North British Railway which was painted all over in Royal Stuart tartan to please Queen Victoria, T. R. Crampton was without honour in British railway circles. For one thing, Gooch's great broad-gauge engines on the Great Western Railway stole their thunder, and when Crampton produced an engine (of "French Crampton" type) called *Liverpool*, which equalled the best of Gooch's in power and speed potential, she simply tore to pieces the much lighter track of the London and North Western Railway which had invested in her. Apart from this one, most of the British Crampton engines were quite deficient in boiler power. Directors were very much afraid of high boiler pressures. On the Great Western, Gooch went his own way in this respect, and being a wise man, did not advertise his pressures to those who, in his opinion, had no business to know about such things.

In America at that time, the Crampton type was likewise a

failure, for the same reasons, only, in respect of track destruction, more so. For John Stevens on the Camden and Amboy Railroad, Isaac Dripps built several, chiefly notable for their 8 ft. driving wheels and for their nightmare appearance. (Some of the French and German engines were of great elegance.) For an example of an American Crampton engine we show the rather more shapely *Lightning*, built by Norris for the Utica and Schenectady Railroad in 1849. She was to be sure much more "English" in appearance and arrangement than other American locomotives of the time, even down to the stuffed leather side buffers, which were very rarely used in the States, even then. She had, however, Norris's bogie, or leading truck. Apparent weak points were in the extreme lightness of the outside frames. She was also much too small, but had she been bigger, it is doubtful that she ever would have held the road at all.

In France, as remarked, the Crampton engine was a great success in the sort of work for which it had been designed; for nearly thirty years it worked all the fastest passenger trains between Paris and Strasbourg, and in the cadets' slang of the Military Academy of St. Cyr, *prendre le Crampton* meant to take the train, to go on leave, to have a night out, long after Crampton locomotives had finally vanished from French railways. Two Crampton engines have survived the years; *Le Continent* as mentioned, and in Germany, the Badenese *Phoenix*, which for many years was kept for instruction in mechanical engineering at Karlsruhe. There is also a beautiful

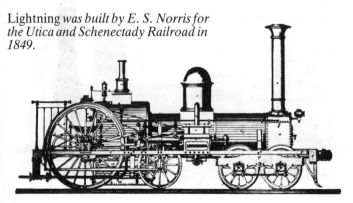

Lightning *was built by E. S. Norris for the Utica and Schenectady Railroad in 1849.*

replica of *Die Pfalz*, one built by Maffei of Munich in Joseph Hall's time for the Palatinate Railway. Both these German Cramptons had Hall's arrangement of cranks, which mounted all the motion-connecting rods and valve gear, outside the external frames, giving splendid accessibility with less likelihood of running hot, at the expense of making the engine somewhat wide. Lineside structures had to be kept at their distance. A solitary English engine with a similar arrangement, by J. E. McConnell of the London and North Western, was so destructive in this way that the men named her *Mac's Mangle*.

In Central Europe, Hall's cranks with outside frames were much used. Maffei built them in engines for the Bavarian State and Bavarian Eastern Railways. Thousands of Austrian locomotives had them; consequently they were to be seen, during the middle and later years of the nineteenth century, fairly continuously on a journey from the French frontier of Baden to the Bosphorus. For an engine of this type and period we show the *Erzsébet* on the Royal Hungarian State Railways. Though not a particularly distinguished example, she was typical of the eighteen-sixties in the Danubian countries and was notable as being the thousandth locomotive built by G. Sigl of Wiener-Neustadt. In the drawing we have shown the ceremonial decorations bestowed on her to celebrate the fact.

Sigl's works produced many bogie engines during this time, unlike most European builders, for the several Austro-Hungarian State Railways and for the Austrian Southern Railway. We show one of these with large driving wheels for passenger train haulage on the more level lines both north and

Sigl's thousandth locomotive.
Erzsebèt *was built in 1870.*

96

south of the Alps. This particular engine dates from the late 'seventies but is best exemplified here. Like *Erzsébet* she typifies the Hall/Haswell locomotive of the Danubian countries, with her deep, slotted, outside frames taking much of her weight and, through the bearings, most of the strains of a heavy engine in motion. The drawings show also how the excentric sheaves and rods were between the bearings and the main cranks. Common French and Italian practice of the time was to mount the valve gear *outside* these, which was T.R. Crampton's arrangement.

PIONEER RAILROADING IN RUSSIA

Neither such arrangement enjoyed any popularity in North America, nor yet in the British Isles, in spite of the fact of Crampton and Hall having been Englishmen and Haswell a Scot. There were many migrant engineers at this time, not least in Russia, which had at first absolutely no mechanical engineering tradition of her own. Harrison and Eastwick from the United States, and Ross Winans also, moved in on the design of Russian locomotives and vehicles. One of Russia's pioneer railroad builders was Major George Washington Whistler, a most memorable American, although to a few he is vaguely known as the father of a great artist and by the rest forgotten either by default, or by design for reasons of political prestige. When he went to Russia, it was virgin railroading land, with the Americans and the British both riding hard for

Sigl's bogie engine for the Semmering Railway; a great prototype.

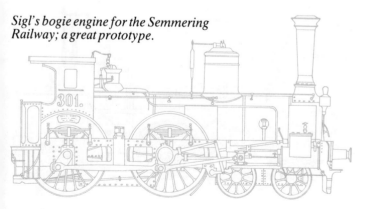

97

the first jump and the Germans watching for the fall of either or both. The British had got in first, as noted before, with the "Patentee" locomotive, though a Uralsk engineer Tcherpinov had produced, as far back as 1833 what appears from a model to have been a "Planet" of sorts, but with the wheels outside instead of within the timber outside frames. It was Whistler who invited the Winans clan and the Eastwick and Harrison outfit to move in on Russia. Whistler was not without cause for honour in his own land; he had built, *inter alia*, the Boston and Albany Railroad over the Berkshires; further when it comes to Russia, he is believed to have been the most interested person present when the Tsar Nicholas I made that alleged ruling in the map with his imperial sword (or ruler?).

Imperial Russian officials had visited the United States and

This Eastwick and Harrison locomotive was built at the Alexandrovsk Works in St. Petersburg in the 1840s.

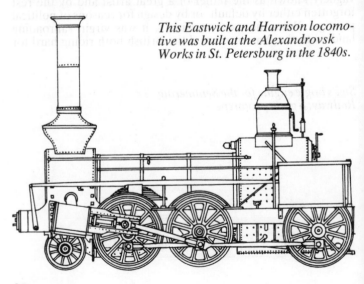

watched the performance of that remarkable *Gowan and Marx* which we have already encountered. Joseph Harrison was the practical engineer of what had become the firm of Eastwick and Harrison. Both partners were invited to St. Petersburg (Leningrad, as we now know it). A contract was drawn up, so handsome that Eastwick and Harrison quitted the United States to set up works in Russia.

During the eighteen-forties, they produced thoroughly businesslike locomotives; four wheels coupled for passenger work and six for freight. We illustrate here one of the freight engines, usually accepted to have been the first 2-6-0 type ever built. The type had been built already for the St. Etienne-Lyons Railway in 1841. In later years, the type was to be very popular, first in the United States and then over most of Western Europe as well as Russia. One of the later American locomotives was named *Mogul* and this became the strict type-name for 2-6-0, or 1-C, engines. In common parlance, it was used for all sorts of large steam locomotives in the United States over many years. Looking at this Russian-American design, we notice several differences from what had already become recognised American practice. She had inside slab or plate frames, like many British engines. The arrangement of the cross-heads and slide bars was also very British. The spark-arrester at the base of the smokestack suggested certain old German designs and, although square in plan instead of round, rather anticipated a form very popular in later years on many railways in Sweden. Contemplating the complete lack of any sort of a cab for the enginemen, one wonders how they survived long hauls across Northern Russia in winter.

This was the first passenger service engine of the Mogul type to be used on the Baltimore and Ohio Railroad. It was designed by John C. Davis in 1875.

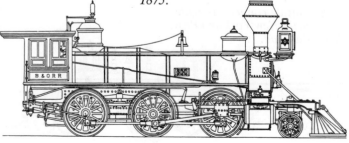

Eastwick and Harrison's first Russian passenger engines were much more in the American style, though the mounting of the valves inside instead of on top of the cylinders was also distinctly European. These old Eastwick and Harrison locomotives were rebuilt in the eighteen-sixties with new boilers and proper cabs (possibly improvised at an earlier date!).

THE CLASSIC AMERICAN

As remarked, the 4-4-0 engine with outside cylinders and a single-pivoted leading truck early became the American national type, though there remained also a vogue in the States for a somewhat similar engine with the cylinder centre-lines much closer together, driving crank axles instead of cranks outside the driving wheels. It was not quite the same as a later arrangement with inside cylinders, for a long time typical of the British Isles, India, parts of Australia and certain Swedish railways such as the Bergslagernas. For a specimen of the old American "inside-connected" locomotive we have chosen the *New York* of the Boston and Providence Railroad. She was delivered to the company by George S. Griggs in December, 1854, and had a remarkable double firebox divided by a longitudinal water-filled partition and fed through two firedoors. Above the twin firebox was a combustion chamber receiving the flames through two short flues in a water-bridge.

The same sort of thing was being tried in England at that time, in the interests of burning coal instead of coke without passing most of it blackly out of the chimney as wasteful and nuisance-valuable smoke. For the rest of this very interesting old locomotive, we should notice the still very short and old-fashioned leading truck, the early form of the chimney

Griggs' inside-connected engine, 1854, on the New York and Providence Railroad.

100

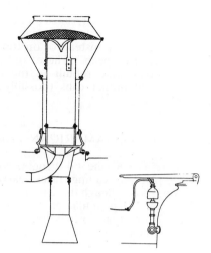

(Above) *O. W. Bayley's double-barrelled firebox was tested on the* New York.

(Right) *Section of the American "diamond-stack".*

(Far right) *This is a typical American-type steam whistle. The cord stretched from the whistle into the driver's cab.*

always known in America as a "diamond stack" and the irregular spacing of the rigid axles on the six-wheel tender. This was quite peculiarly an old New England engine.

Much more typical of the expanding United States was the outside-connected 4-4-0 locomotive with inside link valve-motion, working through rockers the slide valves on top of the cylinders which, in turn, were becoming horizontal instead of inclined.

Thomas Rogers *exemplifies the Classic American type. This was built about 1855.*

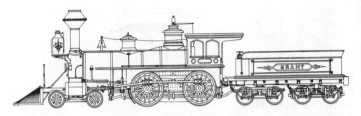

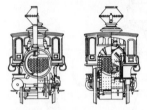

This standard American type was built in 1873 by the Grant Locomotive Works, Paterson, New Jersey.

A classic example of the late eighteen-fifties and, most typically, of the war years in the early 'sixties, is shown in the engine by Thomas Rogers of Paterson, New Jersey. The drawing almost identically represents the "war engine", the *General*, which was built in 1855, and on a rainy day of 1862 was captured by a Northern raiding party and driven on a wild tour of sabotage from near Marietta, Georgia, almost to Chattanooga, Tennessee. The story has often been told, sometimes with doubtful "improvements". The *General*, somewhat altered in later years, and narrowed from the old Southern broad gauge, is still preserved as a treasured relic.

The Rogers design of this time, and many others generally like it, was both simpler and much more accessible than the type exemplified in Griggs' *New York*. A very great improvement was in the long wheelbase of the leading truck, though this still had a rigid central pivot, without any sideplay. The tender was on two generally similar trucks.

Thousands of such engines were built over many years, gradually increasing in size until, at the end of the century, the largest of them were distinctly adequate. We show an example by the Grant Locomotive Works in 1873; a coal-burner with a diamond stack. The enormous balloon stack on the Rogers engine bespoke wood burning. Both engines have the coned gusset-ring to the boiler-barrel, forming what Americans called the "wagon-top" boiler. Such locomotives were to span the Continent. When this was ceremonially done for the first time, at Promontory Point, Utah, on May 10, 1869, the Central

Pacific Railroad's engine from Sacramento was the *Jupiter*, a wood-burner, while the Union Pacific company's No. 119 from Omaha was a coal-burner with an extended smokebox. Otherwise they were very similar engines which the untrained eye might distinguish only by the very differently-shaped smokestacks (the Union Pacific one was straight, with a slightly bulbous cap).

This was the engine of post-war America, as much a part of the American landscape as white wooden frame houses and majestic river steamers. Their enormous headlamps cast night beams far across the prairies, or through vast northern woods, or upon the vertiginous curves of the mountain-rights-of-way. Their music was, to strangers, most melancholy—the soft *choo-choo* of their great stacks, the deep hoots of their whistles and the dolorous clanging of their ornately mounted bells—but to Americans from Atlantic to Pacific these were well-loved sounds, breaking the loneliness of remote places and heralding a nation's advance. By comparison the European locomotive screamed and roared, and to older people especially was often regarded as an intruder. But European youth loved her too, especially in Great Britain where a great coterie of amateurs grew up.

Back to the serious business of design: "Mogul" engines have been mentioned already. The classic form became that shown on p. 103 by an engine built by the Baldwin Locomotive Works about 1870. It is obviously first-cousin to the four-coupled American type but with an extra pair of coupled wheels occupying the space which had previously belonged to the trailing truck wheels, while the four-wheel pivoted truck itself

This Baldwin-built Mogul type is a fine example of the freight locomotive of the 1870s in North America.

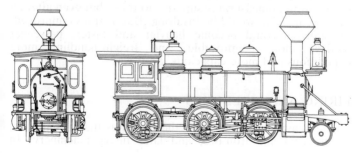

was replaced by a radial one with a single axle—the Bissell truck, for long much favoured in North America. Such locomotives were generally for freight service, but were useful enough for passenger trains in mountain country. It should be remarked that as yet, in spite of many wild claims for record speeds, regular trains were nowhere very fast apart from a few favoured services in Great Britain and France.

For one thing: in most places the track would not stand it. Where rail-ends were simply spiked close together on sleeper or cross-tie, without any proper joint, there was bound to be trouble. The English Great Western company and a few others overcame that at the expense of being over-rigid and difficult to re-lay when that became necessary. The *fish-joint*—the bolting together of rail-ends by lateral plates—was the invention of William Bridges Adams in England. Its use became world-wide, but, Adams being more inventive than businesslike, he never reaped his royalties. After the perfection of Sir Henry Bessemer's process in making steel, by the primary removal of carbon from pig-iron by air blast, and the subsequent re-introduction of a predetermined amount of carbon (*Spiegeleisen*) and ferro-manganese, in 1856, the way was clear for the use of steel instead of iron rails. Two forms of rail were to be used for the best part of a century thereafter; the bull-head rail keyed into iron chairs, much used in Great Britain, India, Western France, the high Alps and for a while in New South Wales, and the flat-footed rail invented by John Stevens in the States, introduced to Europe by Sir Charles Vignoles, which is now generally used throughout the world. Thereafter, up to the use of long welded rail-lengths in our own time, development of the *permanent way* (delightful English expression!) was simply a matter of heavier rails and superior road-bed. Qualities varied from country to country, and from railway to railway. By the end of the century, the Pennsylvania Railroad, the London and North Western Railway and some others were immeasurably superior to the— but even after all these years, that would be insulting! But as tracks improved, locomotives could become heavier and faster, passenger haulage more comfortable, and freight-handling more expeditious and more capacious. So it was.

FIRING AND FIREBOXES

The heaviest haulage, from the first, was that of minerals; chiefly coal and iron-ore, and later heavy machinery. In

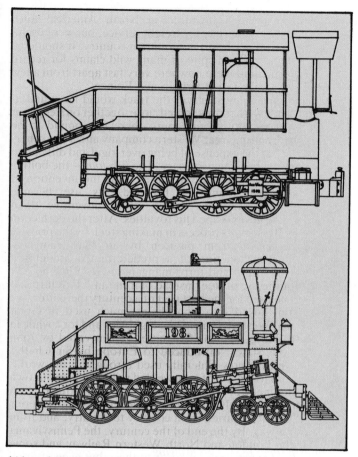

(Above) Camel *was built by Ross
Winans in 1848.*

(Below) *Ten-wheeler designed by the
master of machinery at the Baltimore
and Ohio Railroad, Samuel Hayes.*

coal-bearing country, there was great desirability of the
locomotives burning low-grade anthracite, instead of the
homely cordwood which fed so many American trains, nicely
aromatic and thermally inefficient. England and Scotland, by

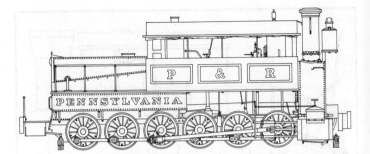

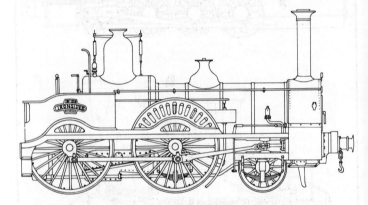

(Above) *James Millholland designed*
Pennsylvania *in 1863.*

(Below) *Joseph Beattie's* Ironsides
*was built in June, 1855. This drawing
shows it after its conversion to
four-coupled wheels. It was in use for
thirty years, being sold in June 1885.*

comparison, were positively luxurious: at first coke, and then
the finest hard steam-coal, were the only things worthy of a
locomotive! England was rich, though headed for a depression
later, and even Scotland was picking-up!

In the United States, Ross Winans was a pioneer in the
economic firing of locomotives, and looking back to the 'forties
we see his once-famous "Camel" type of locomotive (so called

because of its sloping stern containing a long narrow firebox, plus the placing of the engineer in a car on top of the boiler, like an Arab traveller on top of the animal's hump). The first of the Winans "Camels" went into service on the Baltimore and Ohio Railroad in 1848, and was indeed named after that disagreeable beast. Winans built about 200 altogether.

Winans had Confederate sympathies in the Civil War, but he had already closed his works in Baltimore. Like Eastwick and Harrison, he found another good market in Russia, where, more than locomotive work, he was strongly to influence car design. In America, however, the Winans type was quickly revived by Samuel Hayes, who most importantly suppressed the leading coupled axle and added a leading four-wheel truck, centred under the cylinders as in the classic American Type. We show a Hayes "Camel" of 1854. Hayes was Master of Machinery on the Baltimore and Ohio Railroad from late in 1851 to the spring of 1856, and the last of his engines, of this type, ran until 1901, after which it was preserved at Purdue University.

A very eminent American designer of those days was James Millholland, who in extreme youth had helped with the building of Peter Cooper's *Tom Thumb*. He became Master of Machinery on the Philadelphia and Reading Railroad in 1848, and almost at once applied himself to getting the last ounce of steam out of low-grade anthracite, in a capacious—and ultimately very wide—firebox whose final form is generally known as the Wootten firebox, named after the President of the "Philly" (the P. and R. Railroad). Millholland's express passenger engines such as *Hiawatha*, *Minnehaha* and *Kosciusco* were famous in their day, and of their kind very beautiful and unusual examples of the usual American type. We illustrate here his *Pennsylvania* of 1863, the first locomotive to have twelve coupled axles. It was in many ways Winans' *Camel* greatly enlarged and modernized by the standards of the time.

In England at this time, even using such beautiful steam coal as that of South Wales, there were problems about any sort of coal burning. Americans might watch affectionately a belching of thick black smoke from the advancing stack, but in all Great Britain there were not entirely-dormant penalties for such emission. Yet coke was abominably expensive, and in any case, no locomotive engineer there liked to see all that unburnt stuff being wasted on the fresh air. Joseph Beattie, an ingenious Irishman on the London and South Western Railway, James McConnell, another Celt on the London and North Western, and James L'Anson Cudworth, a northern Englishman on the

Prins August *was the first passenger locomotive used by the Swedish State Railways, 1856.*

South Eastern Railway, all set about securing complete combustion by elaborate double firebox arrangements, with midfeathers and water-bridges—and in Beattie's case thermic syphons and complex combustion chambers. We have already seen something of the kind in the little *New York* of the Boston and Providence Railroad. Beside some of Beattie's weird arrangements, her firebox was a simple kitchen kettle!

For the record we show here an outline of Beattie's London and South Western locomotive *Ironsides*, built in 1855. Very briefly she had a single driving axle, but a coupled trailer was quickly incorporated. The drawing shows the engine as she was in 1881, for she had a long life of thirty years. Her firebox was divided into front and back compartments, divided by a water-filled half arch. Heavy firing was in the rear part, the forward fire being kept as far as possible incandescent all the time. Further, at one stage, there was a combustion chamber in the boiler barrel, containing a thermic syphon, so at the expense of some fairly heavy maintenance, a Beattie boiler could be relied on to get very much more than a pound of steam out of an ounce of coal, and so it did. A century ago, these engines—for many were built—were not to be beaten at this game. Mechanically, the design was one schemed by Sir John Hawkshaw in the 'forties. For years it was very popular in Europe, and steady too, in spite of all that overhang at the front. The first locomotives of the Swedish State Railways in 1856, built by Beyer Peacock and Company of Manchester, were of this type, though without Joseph Beattie's weird and wondrous fireboxes which probably lost on maintenance what they won on fuel economy, but certainly made their designer nicely rich by patent royalties. The Netherlands State Railways Company (lovely mixture of interests!) had some very fine examples in the late 'seventies, and so did certain Prussian lines such as the Bergisch-Mârkische Eisenbahn. On the latter, be it remarked here, a Wagner other than the one who composed *Den Ring* mounted water purifiers in an extra dome on the boiler, the feedwater being allowed in this dome to drop through a series of trays and so, under steam temperature, to deposit its mineral content before this could settle on the boiler plates and do cumulative mischief as the boiler grew older.

All these were extremely elegant engines in the European style; they seemed to breast their head-winds like great swans rising, though to American eyes, in absence of that leading truck, they seemed likely to become airborne in a less desirable

Jenny Lind, *designed by David Joy and built by Leeds Engine Foundry in 1847.*

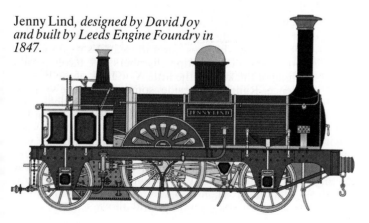

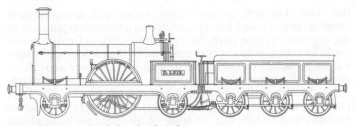

Dom Luiz *was built for the South-eastern Railway of Portugal by Beyer Peacock of Manchester in 1862.*

way than that of the noble bird. They were in fact remarkably steady, and fast too. On the London and South Western they could *average* their mile-a-minute, *by schedule*, with a favourable gradient.

Another archaic type of British locomotive which lasted for a very long time and was to be used in many other countries was one which had originated in the late 'forties. It was generally known as the "Jenny Lind" type, a famous early example having been named after that sweet singer from Sweden who, incidentally, was to number among her later lovers one James Staats Forbes, for many years Managing Director of the London Chatham and Dover Railway. The drawing is reasonably explanatory. The type was describable as a "Patentee" with only inside bearings to the driving axle. Thanks partly to increased boiler pressure, "Jenny" was a very successful engine. A famous later example of the type is the *Dom Luiz*, built by Beyer, Peacock and Company of Manchester for the South Eastern Railway of Portugal in 1862, which has survived as a beautiful relic. Her domeless boiler (which remained unchanged though doubtless renewed) exemplified an old argument as to whether a dome really provided valuable steam space to the boilers as well as a convenient place to house the regulator and steam-pipe opening, or whether it merely weakened the boiler shell. The argument never was resolved. Looking back to Gooch's *Great Britain* on the Great Western Railway, and to certain very efficient little engines known as the "Eddy Clocks" in New England, and forward to the last steam engines built by that same Great Western, and to the last giant steam engines built for the Canadian Pacific Railway, we note that all these were domeless, while most engines in the rest of the world had as much dome as the loading gauge would allow.

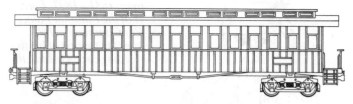

This 1865 American car was in general use on most railroads. It was 50 feet in length and weighed just under 16 tons.

Another curiosity! Carl Friedrich Beyer, joint founder of the Manchester firm, was a German from Plauen in Saxony. His migration is comparable to the nearly opposite one of Haswell to Vienna. Their designs were to be regarded, respectively, as typically British and typically Central European.

CAR DEVELOPMENT

At this point, motive power can rest awhile, and as freight and mineral handling was as yet in fairly primitive vehicles, let us turn to passengers. Their conveyance across the North American Continent naturally demanded something fairly superior to carry them, and there were similar, if more regional, demands from Biscay to the Urals. Two starting points: America tended to base her passenger cars on the canal boats which had preceded them; Europe still stuck to the stage-coach idea, with increasing modifications.

America produced the first really spectacular improvements. The long "shabby omnibus" which had scarcely pleased Charles Dickens became gradually refined. People might still use its central passage as what someone called an *elongated spittoon*, but it had many good points. Its little two-by-two seats with reversible backs were adequate enough by day. A pot-bellied stove kept people warm in winter. Near this was a little annexe with a bottomless can to serve natural calls. There were candle-lanterns at night. In the middle 'sixties these were replaced by quite good kerosene lamps with Argand burners, slung in clerestory decks which not only raised them above the passengers' heads but gave better daytime lighting and better ventilation at all times, as well as rather more head-room than the old, nearly flat, coach roof. Such was the dust on old-time

111

George Mortimer Pullman, 1831 – 1897. Founder of the Pullman Palace Car Company in 1867 after the success of his first sleeping cars. He built the Pioneer *in 1864.*

journeys that people thought twice about opening the windows even in summer even if they could. *Sticks like a car window!* ran an ancient American advertisement for some adhesive. By that time, such windows were slid upwards into a hollow top quarter in America, and lowered ("drop-lights") into the hollow of the primeval carriage door of Western Europe.

Here we show a drawing of a characteristic American railroad car of 1865. It was quite spacious, though cramped as to seating, giving a foretaste of large passenger aircraft just a century later. There were small seats with reversible backs, often quite parlorly trimmed in the best available red plush. In the words of Mark Twain: *Uncomfortable, but stylish*.

Travel in such a car could be tolerable enough in the daytime, but it was penitential at night. Many attempts, as briefly noted, had been made to produce a reasonable sleeping car. George Mortimer Pullman, a cabinet-maker, had the idea of making the daytime seats pull out flat to meet in the foot-space, while folding berths were brought down from the ceilings each side of the aisle, beds being made up on both levels. During 1858-59, he converted three ordinary day cars of the Chicago and Alton Railroad. They were a success, but war interfered with his plans and, not being interested in the hostilities, he went west until the four awful years were over. In 1865, he produced the *Pioneer*, the first real sleeping car. The roof was altogether higher, with a liberal clerestory in the middle, and sliding boards prevented the recumbent passengers from kicking each other's heads. It was an instant success and became an American prototype that was to be followed for several generations. The classic Pullman sleeper had arrived, and though funny stories have been told about the business of undressing in those berths behind buttoned curtains, and

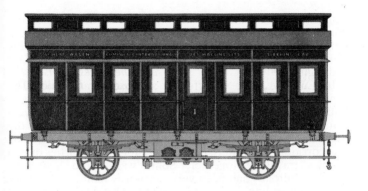

This Austrian sleeping car was built by Hernalser and Company of Vienna and exhibited at the Vienna Exhibition in 1873.

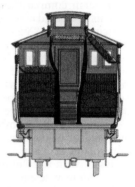

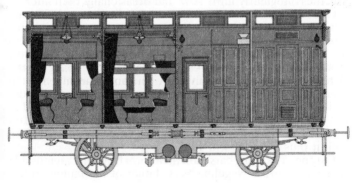

embarrassing mistakes (the main portion of the car formed a dormitory for two-dozen assorted men and women), the Pullman sleeper could be adequate enough, and reasonably comfortable, even on a transcontinental journey of several days and nights.

In Europe, quite a different approach was made. The compartment arrangement was kept. (For that matter, a Pullman sleeper would offer, at a price, a couple of private compartments, each with two berths, at one end.) In the 'sixties, Russia provided four-berth compartments reached by side corridor, which were nice enough for family parties or two couples travelling together. Some were quite luxurious. Russian railroads had been very spaciously built, unlike those of the pioneer lines in Great Britain (which paid the penalty of so many pioneers). A Russian sleeper of 1867 had five four-berth compartments, a small middle saloon, and above this an overhead observation compartment reached by a stair, anticipating the "Vista Dome" of America, some ninety years after. The arrangement made an intermediate appearance on the Canadian Pacific Railway in the early nineteen-hundreds. For the rest of this Russian sleeper, it was entered by open platforms each end, as in America at that time, and had some sort of water-closet beside each entrance. The car has been several times illustrated, even recently.

Less familiar to posterity has been the neat little Austrian sleeping car shown on p. 113, which combined the compartment idea favoured by Georges Nagelmackers, a Belgian, with the Pullman arrangement of seats and berths. It was evidently designed for international travel as from Vienna to Berlin or Munich. Something rather like it was to be found in Scotland, on the slow night trains between Glasgow and Inverness, regularly until 1907 and occasionally thereafter.

Our drawing of the Austrian car is quite sufficiently self-explanatory. One notes the under-floor stoves for heating through gratings. Steam heating from the locomotive was still in the academic stage, which means that nobody had ever seen—or felt—it. In the Austrian sleeper, the shape of the clerestory deck was markedly un-American. Its arrangement of windows and ventilators followed the convention of the side- and door-windows of the old English side-door coach, and in this form it was soon to become very popular in Prussia and on certain English railways, notably the Great Western, which built it thus from 1874 to 1903.

European sleeping cars took their form from that memorable Belgian, Georges Nagelmackers, father of the famous Inter-

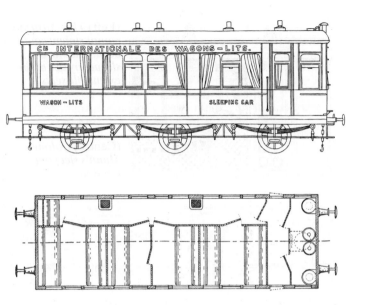

Georges Nagelmackers' six-wheel car,
shown at the Paris Exhibition of 1868.

national Sleeping Car Company. After some struggling, he teamed-up with an ingenious if rascally American, Colonel William D'Alton Mann, who had got hold of some British capital. The Compagnie Internationale des Wagons-Lits became, briefly, Mann's Railway Sleeping Carriage Company, Limited. The Nagelmackers cars were divided into compartments with transverse berths and short side corridors leading to lavatories. Colonel Mann called them *boudoirs*.

Supplementary to sleeping berths giving passengers a fully recumbent position, there was a limited vogue in Central and Eastern Europe for a carriage furnished with *chaises-longues*. The accompanying sketch (p. 116) shows what was available in the eighteen-sixties to those willing to pay for it on such various lines as those between Vienna and Trieste (this example) and between Kiev and Odessa. It was certainly better than no sleeping berth at all, and less disquieting than certain coffin-like berths of the pre-Pullman period in North America. The specimen shown was at one end of an otherwise ordinary first-class coach, and was approached by an end-balcony which also gave access to a very small water-closet. A stove with a

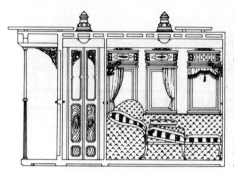

(Left) *This Austrian compartment could be turned into a sleeper by use of the extensible seats.*

(Centre) *Colonel Mann's sleeping boudoir.*

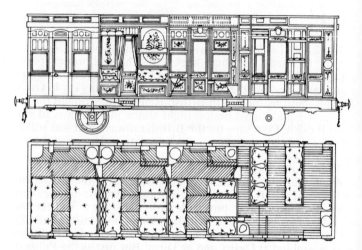

(Below) *Fraser's broad-gauge family saloon. Great Western Railway, 1866.*

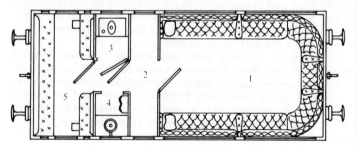

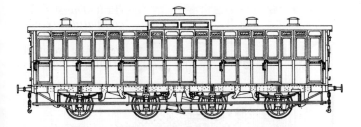

This second-class coach was built for the New South Wales Government in 1867.

primitive vapour-heater was mounted under the carriage frame.

French designers at first favoured a sleeping berth formed by tipping forward a very high-back seat. The seat proper folded underneath while the reverse side of its upholstered back, now horizontal, became the bed. The first real sleeping car in Great Britain, built by the North British Railway in Scotland for the Glasgow-Edinburgh-London night trains in 1873 was of this sort. It did not last long on the British railways, but was still to be seen in France in the nineteen-twenties. The official term was *lits-salon*.

From America, George Mortimer Pullman made a move towards European users, beginning with the Midland Railway in England (1874). The cars were of pure American type, very gorgeously decorated, but made rather smaller to suit the the railway industry. Several British railways invested in Pullman sleepers, and so did Italy. They were prefabricated in the States and assembled at Derby and at Turin. But just as the Pullman became the standard American form, so did the Nagelmackers type become characteristic of Europe. Colonel Mann certainly made a few cars in his own country, but his chief interests there seem to have been in blackmail, in conjunction with a scandalous newspaper, which made him a lot of money. He really *was* a colonel, be it added!

Apart from sleeping cars and special cars for eminent personages, the West European coach remained in its classic side-door-compartment form. Australian railways, unlike those of Canada, were very English in style, and we show a second-class carriage with a central guard's-brake compartment, built for the New South Wales Government Railway in 1867. It was notable, for a British carriage of that time, in

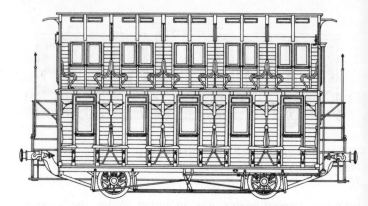

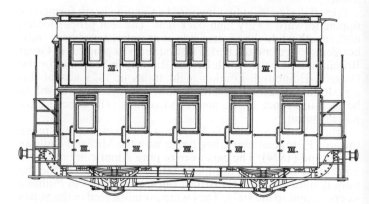

*An Austrian third-class double decked
car.*

having eight wheels, the axles having a certain amount of
side-play though no bogies were employed. As a coach, it was a
comfortless thing; its very narrow compartments had
low-backed wooden seats. The guard had a view either way
along the train from the raised lantern roof, and could also keep
an eye on his passengers through two peep-holes.

Already by the eighteen-sixties, there was heavy movement
by railway, of daily workers in and out of the greater cities, both

labourers and more and more office workers, the former at special very-cheap rates. Their trains grew longer and longer; too long for many of the stations in areas already built-up. If possible then, the passengers must be carried on two decks. From very early days, certain French railways had mounted roof seats on carriages and so had certain pioneer American lines, as we have seen.

At the Vienna Exhibition of 1873 there was shown—along with the Hernalser sleeping carriage already noticed—the double-deck third-class carriage now illustrated. Really, it was a masterpiece of reasonably liberal accommodation within a short body, seating fifty persons in the lower compartments and forty in the gang-wayed upstairs portion. The latter had a clerestory, perhaps against claustrophobia, though this feature was lacking in similar vehicles which subsequently appeared, in large numbers, on the lines of Paris, Berlin and Copenhagen. The well-construction of the iron underframes will be noted, allowing for the maximum possible head-room in the passenger space. Again, England's unfortunately small tunnels made such a vehicle impracticable in London, where it would have been very useful, but even during the unpleasant nineteen-forties, Paris could muster suburban cars very similar to this old Austrian specimen, on both the Eastern and the Western lines. On the latter, too, a more disagreeable sort with open sides to the top deck lasted for nearly as long. *Impériale* was the French term. They were terrible things in cold weather, and dangerous too: also nasty when there happened to be some drunk and disorderly character on the top deck. Spain still has some old double-deckers at the time of writing.

From these Spartan vehicles, let us turn to a much more distinguished sort. Royal and other great personages had commanded luxurious railway carriages from the time when Queen Victoria made her first English journeys by rail in 1842. She, and the Russian, French and Austrian Emperors, all had sumptuous vehicles built for their long journeys, and so did the numerous German kings and many others. Many of these vehicles have survived the years in European museums. Very noteworthy examples are those of Queen Victoria, built by the London and North Western Railway in 1869 and now at Clapham Museum, and of Maximilian II of Bavaria, now at Nuremberg. This, somewhat older than Victoria's, is an almost incredible specimen of ornate Wittelsbach-Rococo, outside as well as within. It is more generally associated with King Max's successor, Ludwig II ("Mad Ludwig" to people who were not Bavarians) whom it must have suited admirably.

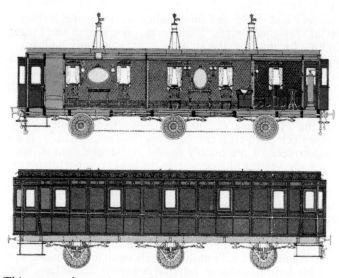

This extremely spacious and luxurious car was built for the Russian tsar. It was used on the Moscow-Kursk Railway.

These two, as remarked, may still be looked at, and marvelled at; and they have been fairly often illustrated. More likely, therefore, to satisfy retrospective curiosity is the example on p. 120, which was made for the Tsar.

The carriage was extremely spacious and most luxuriously arranged. Its flat roof and consequent lack of headroom may have helped to solve the problem of keeping Imperial Majesty warm on long winter journeys. The body was very efficiently insulated and heavily upholstered to the same end. The closed receivers to the lavatory fixtures will be noted. Open drains could let in the intense cold, or alternatively ice-up. (An unhappy experience of Queen Victoria on a wintry journey from Scotland to the South. Everybody suffered!) The Russian arrangements curiously anticipated those of the passenger aircraft in our time.

That will do for cars, for the time being. The worst of them were very awful indeed, but the best were handsome, elegant, and even luxurious. It depended on one's purse or bank.

George Forrester of Liverpool designed and built this 2-2-0 outside cylinder engine, Vauxhall, *for the Dublin and Kingstown Railway in 1834.*

Velocipede, *Alexander Allan's 2-2-2 passenger locomotive for the London and North Western Railway 1847.*

PROGRESS IN DESIGN

Reverting to traction, the coupling of driving wheels on locomotives was fairly general by 1870, even on the fastest express passenger service, but two notable exceptions were Great Britain and France, on certain railways having a demand

121

for fairly light, high-speed trains making but limited stops. We have not yet noticed a sort of locomotive known in England as the Allan type, and in France as *Le Buddicom*. (Allan, incidentally, was a Scotsman and Buddicom an Englishman!) Its origins were ancient. The *Vauxhall* of the Dublin and Kingstown Railway, here shown, was built by C. and S. Forrester in 1834. Noteworthy are the outside cylinders and drive, the former supported by the inner and outer frames with slots in the latter over the piston rods and cross-heads. Otherwise the engine was a "Planet" with inside bearings only. At the end of the 'thirties, the Grand Junction Railway (Birmingham to Liverpool and Manchester) was having terrible trouble through the breakage of the early crank axles. Conversion of the engines to an outside-cylinder arrangement was rapidly carried out on Forrester's plan. The credit has been given variously to Joseph Locke (builder of the line), to Alexander Allan, to W. B. Buddicom, and to Francis Trevithick who was in charge of the locomotives for the railway company. Anyway, they lengthened the Forrester design, which had been Allan's work, into a six-wheel engine. All these men were closely associated at the time. Locke, a civil engineer, went to France to build the railway from Paris to Rouen, Dieppe and Le Havre. Buddicom went also, to build engines at Rouen. New engines of the Allan-Buddicom type appeared almost simultaneously on what became respectively the Western Railway of France and the London and North Western Railway. The first in fact appeared at Rouen in 1844, and early in 1845 the *Columbine* was the first locomotive to be built at Crewe in the English Midlands, one of the world's most famous locomotive towns. Both engines worked into the present century, and both *Columbine* and one of the Frenchmen are still in existence and capable of being steamed.

Our present drawing of this type shows the *Velocipede*, built at Crewe in 1847, and is based on one of Alexander Allan's own drawings. She had 7 ft. driving wheels and the high pressure, for those days, of 120 lb./sq. in. This type of engine was very widely used. It abounded in Scotland, whither Allan later returned, on the North Western line in England, on the Western Railway of France, and for a while in Canada. Early examples were to be found on the Spanish Barcelona-Mataro Railway, on the Roman Railways in pre-Victor-Emanuel Italy, and in India. The Great Western of Canada and the Grand Trunk Railway in the same country added leading four-wheel bogies to allow for the usual North American track standards of the period, and this feature, already adopted by the Tudela and Bilbao Railway

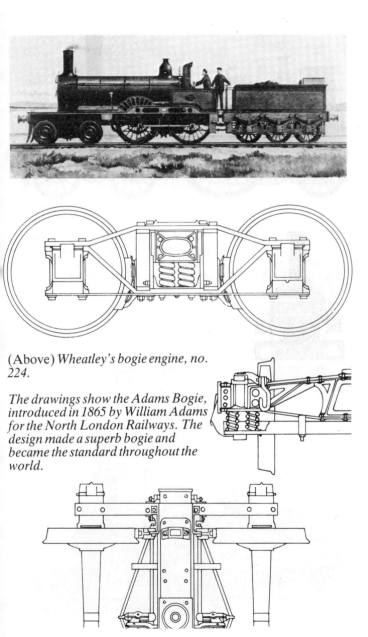

(Above) *Wheatley's bogie engine, no. 224.*

The drawings show the Adams Bogie, introduced in 1865 by William Adams for the North London Railways. The design made a superb bogie and became the standard throughout the world.

Patrick Stirling designed this lovely engine for the Great Northern Railway in 1870. She had a 5 ft. 6 in. firebox and 4 ft. 1 in. trailing wheels.

(Below) *The* Batilly, *a 2-4-0 of the 774 series, was built in 1870 for the Compagnie de l'Ouest, France.*

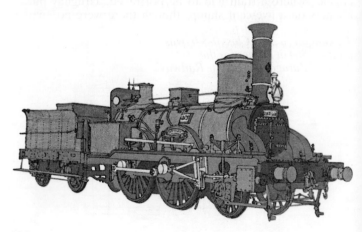

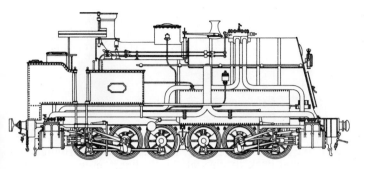

Petiet built this tank-engine for the
Chemin de Fer du Nord in 1863.

in Spain, was adopted by David Jones on the Highland Railway, in Scotland, a steep and curvaceous line, from 1873 onwards. By that time, the engines almost invariably had coupled wheels. Jones of the Highland built the type, very much enlarged, right up to 1892. Survivors of these were still at work in 1930, so the type lasted the best part of a century.

A very famous British design, with single driving wheels 8 ft. in diameter, was that of the Scots engineer Patrick Stirling for the Great Northern Railway, 1870; so famous indeed that artists of last century were inclined to use it as a "stock railway engine" where a train had to be portrayed. Uruguay put its likeness on a five-cent stamp, though there were no engines

This engine, a 2031 Class 0-8-0, was
built in 1864 in France for the
Madrid-Zaragoza-Alicante Railway.

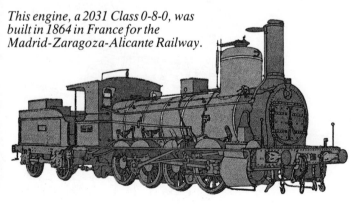

Jules Petiet, Chief Engineer of the Chemin de Fer du Nord, 1845 – 1872.

much like this at Montevideo. The design had forerunners, the oldest going back to days at the Stirling family's foundry at Dundee in the eighteen-thirties. A rigid-pin bogie supported the front end. The idea was not that of following curves (the rigid pin was against that) but, to use Stirling's own expression, of *rolling out the road*. They were indeed very steady engines; fast, too, and more powerful than their long-legged aspect

In 1870, Nicholas Riggenbach constructed this rack locomotive.
1 Driving axle
2 Crank shaft

3 Clearance between the driving cogged wheel and the rack
4 Safety arrangement against derailment

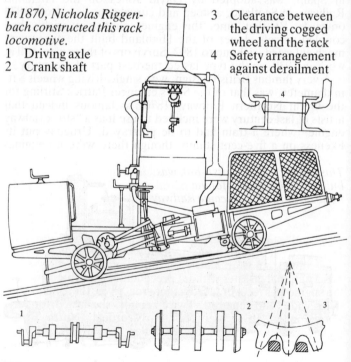

might suggest to strangers. The class was built over twenty years to the beginning of the 'nineties, with very little alteration. During intense competitive running on the East and West Coast routes from London to Scotland, during 1888 and 1895, such engines made start-to-stop average speeds exceeding sixty miles an hour from London to Grantham and thence to York. Probably their best run was one of 181 minutes for the 188.25 miles, with a four minute stop for changing engines at Grantham. That was in 1895, when the design was already half a century old. A typical load was of two vans or baggage cars, three coaches and a sleeping car.

But what was to be a highly characteristic passenger locomotive on British lines for the best part of half a century, in despite of a liking for single driving wheels on express engines, was that with four coupled wheels, a leading bogie with controlled sideplay to the pin, inside frames and inside cylinders. It really belongs to our next period, but we show here North British Railway No. 224, built in 1871, designed by Thomas Wheatley (a rare specimen of an English locomotive engineer serving a Scottish railway; David Jones of the Highland Railway was another). The bogie has been mentioned. What was to become the standard form of the world had been invented by William Adams of the North London Railway (sometime of Victor-Emanuel's Sardinian Navy) in the late 'sixties. Lateral movement of the pivot was controlled in early examples by inclined planes and later by springs on each side of a slot in which the central pin moved. It was a perfect "truck" in American parlance, both steadying the engine and making her amenable to quite sharp curves at high speed.

North British engine No. 224 must have been one of the very first real express engines to have such an appliance. It must be admitted that she was, compared with many in Europe and America by then, somewhat undersized, but many pioneer designs have been modest enough. She had a frightful misfortune in 1879 when she went down, with the northbound mail train, in the collapse of the first Tay Bridge on December 28. There were about seventy-eight persons involved when bridge and train were blown over together. None survived. It was a national disaster without precedent. After three months, the engine was recovered from that great river, there nearly two miles wide, and repaired. Twice rebuilt, she lasted for forty years longer. In view of her early ordeal, North British enginemen rather callously called her "The Diver", a name which stuck.

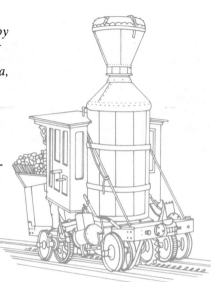

(Opposite) *Detail of*
Milan Railway Station *by
Angelo Morbelli, 1863 –
1919. Courtesy of the
Galleria d'Arte Moderna,
Milano.*

Old Peppersass *on the
rack railway on Mount
Washington, 1869. The
line had a maximum gra-
dient of 1 in 3.*

Turning to less orthodox types of locomotive, it was not only
on the Semmering that there was interest in engines with paired
motor units, as in Cockerill's *Seraing* and the *Wiener-Neustadt.*
In France, there was Jules Petiet, a very memorable locomotive
man whose reputation abroad suffered from British and
German indifference and, one regrets to say, public sneers in
America. We show one of a series of tank engines for heavy
freight, built for the Northern Railway of France in 1863, with
four cylinders, paired at opposite ends of the frame. The lateral
bending of the two sets of wheels, which we might term
motor-units, was allowed for by the use of Beugniot's balancing
levers, with the very long copper steam-pipes and exhaust pipes
running fore-and-aft from the locomotive's mechanical
fulcrum. But more interesting to the eyes of later years was the
apparatus on top. This combined the functions of feedwater-
heater and steam collector and regulator box of the sort T. R.
Crampton had already introduced to French railroads. He had
first made this arrangement in rebuilding one of Cail's
Crampton locomotives on the same railway. Thus equipped,
the transformed "Crampton" resembled nothing else on earth,
but the economiser evidently worked. Petiet built some
passenger engines arranged in much the same way, but

129

with a set of cylinders and single driving axles at each end, separated by three carrying axles. These too were tank engines, without separate tenders, and were perhaps the most fantastic-looking locomotives in regular public traffic for many a long year. One of them, we were told by the late J. C. Cosgrave, a very senior colleague, took the last train out of Paris before the siege of 1870.

Steep-grade mountain railways made their first appearance in the eighteen-sixties, almost simultaneously in America and Europe. Rack and pinion propulsion, as we have seen, was one of the oldest things in steam traction, but its use had long been dormant. In the States, Sylvester Marsh made a model locomotive for very steep grades, using a geared engine and a pinion wheel engaged in a ladder-shaped rack between the running rails. Nicholas Riggenbach of the Swiss Central Railway patented a very similar form in 1863 and applied it on the Kahlenberg Railway, Austria, in 1866. In that year, Marsh gave a similar demonstration in the States. Both were successful, and both resulted in the building of full-scale mountain railways. In 1867, construction began on the Mount Washington Cog Railway in New Hampshire. It was opened to the summit in the summer of 1869, preceding by two years the partial opening of the Swiss line from Vitznau to the summit of the Rigi, as far as Staffelhohe. Rigi-Kulm was reached in July, 1873. Riggenbach's first Rigi locomotive, named *Stadt Luzern*, and Marsh's, which went down in history as "Old Peppersass", both had vertical boilers, hence the skittish nickname of the American engine, whose boiler reminded people of a savoury-sauce bottle. Later boilers on both lines were of ordinary locomotive type, but at an angle to the frames to suit water-level on the very steep gradient. "Old Peppersass" was preserved, but a well-meaning celebration trip comparatively recently ended in a sad and very destructive accident. Portions of the engine were retrieved. A Vitznau-Rigi locomotive, in original form, is preserved in the famous Verkehrshaus, Lucerne. These two lines were the real pioneers of steep-grade tourist railways to the tops of popular mountains in holiday country. The maximum gradient on the Rigi line was 1 in 4. Both lines are still doing well. The Vitznau-Rigi line has long been electrified.

James Barraclough Fell, an Englishman, used instead of rack-and-pinion a central rail laid on its side and gripped by horizontal driving wheels on the locomotive. A line on this system was built on Napoleon's military road over the Mont Cenis pass between France and Italy in 1867, with a

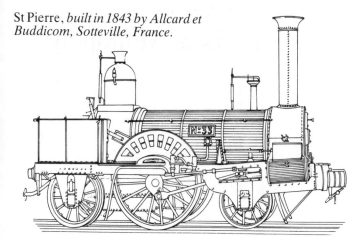

St Pierre, *built in 1843 by Allcard et Buddicom, Sotteville, France.*

life-expectation of a quarter-century while a tunnel was being built under the Col de Frejus, but mechanical advance dashed people's hopes. Thanks to Sommeiller's compressed-air rock drills, the tunnel was ready by the autumn of 1871 instead of the early 'nineties. In those few years, however, the Fell railway carried passengers and also the Indian Mail from London to Brindisi. Fell was a disagreeable character, so one feels less sorry about his disappointment. Even the "Fell System" went back to a joint patent of Vignoles and Ericsson, respectively masters of track techniques and mechanical invention. It was used in France on the Puy de Dôme Railway.

A virtue of both the rack and the centre-rail was that they made extremely efficient braking systems possible at a time when this was one of the weakest points of railway mechanical engineering. The most celebrated Fell line was that of Rimutaka in New Zealand (1885-1955) which also, ultimately, was replaced by a great tunnel. The arrangement is still used for braking on the Snaefell Mountain Line in the Isle of Man.

In 1863, the world's first underground city railway was opened in London, between Paddington in the west and Farringdon Street in the city proper. The line was first built on broad gauge (7 ft.) and briefly worked by the Great Western company. For it, Daniel Gooch built locomotives with surface condensers, the steam being turned into the water-tanks during passage underground while the smoke was kept as clean as possible and left to look after itself.

Chapter 5
The Years of Monopoly

In the last quarter of the nineteenth century, many people believed that the steam railway train had reached perfection. It came to span the world in its journeys. One could live in a train for several days and nights and be none the worse. The train handled freight by land on a scale which even its pioneers had scarcely foreseen. By 1900, locomotives were already so large that many considered that they had reached their limit in size. Electric traction and internal-combustion engines were both over the horizon, though pooh-poohed by the old school of engineers.

Still the steam locomotive maintained its classic form and certain recognized varieties, one of the commonest of which was that faithful old "American Type", the 4-4-0 engine with outside cylinders supported by a bogie. In Europe it almost invariably had plate- or slab-frames, first applied to the type in Eastwick and Harrison's locomotives for Russia and by Haswell in Vienna (1844). We show two beautiful and efficient examples with much in common, though serving very different country. The Finland State Railways' No. 11 was one of a series built in Scotland by Dübs and Company of Glasgow as far back as 1869. More were built by Sigl in Austria in 1875-76 and, with little variation, again by Dübs in 1893 and 1898, and by the Swiss Locomotive Works, Winterthur, also in the latter year. Twenty were still running in 1927, the oldest dating back to 1875 (five Sigl engines). The design was entirely that of the Scottish firm, and in early days the engines were wood-burners, with the Russian form of spark-arresting smokestack. There was a reasonably liberal cab for the enginemen, who needed it in Northern Europe!

Speeds were low in Finland, as in the old Russian Empire generally. The engines were both strong and flexible, with the Adams bogie, and their long life is a sufficient certificate of their usefulness.

This type—the slab-framed "American" which America proper scarcely knew—came to abound in many places, especially those under powerful British influence in railways such as the South American republics, and to some extent in China. Austria, Prussia and Italy all built it; to it belonged the handsome Class CC on the Swedish State Railways, which last had a remarkable bogie arrangement.

For an English example we show No. 471 of the London and South Western Railway, designed by that William Adams who produced the modern locomotive bogie. This particular engine, built in 1884, was one of many generally similar and built over the years 1880-87. They were strong, powerful and fast. In their prime they competed with the mighty Great Western Railway, then still on broad gauge. Over forty years later, we had a sprightly run with this very engine, No. 471, following a late arrival at Okehampton in the West of England. The colour scheme shown is the original one; later greens were much lighter, right up to the end in the middle nineteen-twenties.

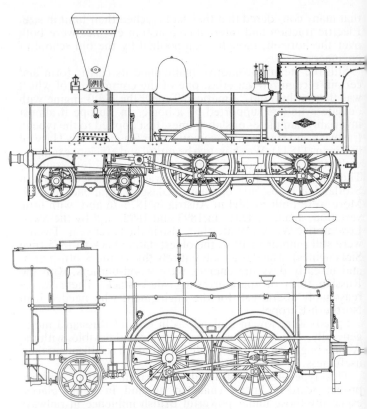

(Above) *Finland State Railways No.
11, built by Dübs of Glasgow in 1869.*
(Centre) *Passenger locomotive for the
Carl-Ludwigsbahn, built by Kessler of
Esslingen.*

About contemporary with these were some express
passenger locomotives which to American eyes were almost
incredible, for they had the coupled wheels on the leading axle;
no bogie—no other axle in front of it. The type had originated
for freight haulage right back in the beginning. The English
variety had inside cylinders; the old Scottish Kessler of
Esslingen produced a large-wheeled express engine of the latter
sort for the Galician Carl-Ludwig Railway in the north-eastern
part of the old Austrian Empire. Hartmann of Chemnitz

built the type for Spain. One doubts that any really high speeds were attempted. The outside reciprocating masses, while all right on an old Scottish coal train, would have made a fast-running locomotive very unsteady. The arrangement allowed, however, for a good long firebox extending well back into the cab. One must say that it was rather a courageous design on the part of Kesslers!

The English version with inside cylinders, designed by William Stroudley for the London, Brighton and South Coast Railway, is exemplified by *Gladstone*, first of a series of thirty-six built at Brighton from 1882 to 1891. Six very similar engines were older (1878-80). The "Gladstones"—though Americans might shudder to look at them—were not only extraordinarily powerful for their modest size; they were perfectly steady at seventy-five miles an hour, and for many years they worked the heaviest Pullman car trains between London and the South Coast of England. *Gladstone* herself is lovingly preserved, and the last of the class to remain in service steamed until the autumn of 1933. In both design and workmanship they were superb examples of old English practice.

No other railway dared to build fast passenger locomotives with 6 ft. 6 in. coupled wheels leading, but only once was a "Gladstone" in a serious derailment, and that was through failure of an old iron bridge, and no engine fault. The Northern Railway of France tried the type as an experiment, but no more, subsequently substituting a four-wheel truck for the leading big wheels. Its track was below English standards in those days, and there may have been trouble.

LOCOMOTIVE ACCESSORIES

Some notes are needed on locomotive accessories. Evolution of valve-gear makes a fascinating study and has been the subject of various learned papers. As we have seen, the expansive use of steam dated back to the days of Gooch and Robert Stephenson's firm in the eighteen-forties, while a primitive form of expansive motion had been used even on the American experimental engine by James in 1831. The object of these new gears, apart from admission, exhaust and reversal, was to make all the valve events—cut-off, exhaust, compression and admission—occur earlier in the cycle by *linking up*. Through use of the link (curved in Gooch's and Stephenson's, straight in Alexander Allan's), moving the reversing lever nearer

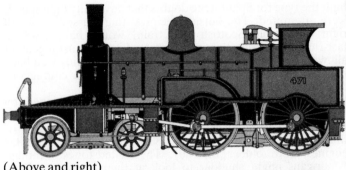

(Above and right)
*No. 471, built for
the London and
South Western
Railway in 1884.*
(Centre) Glad-
stone, *built in
1882.*

to its central (neutral) position, steam could be economized
while running fast. From the early use of these gears dated the
first real express-running over long distances. With the old
non-expansive gears, engines would have lost their
steam.

Radial valve gears, ultimately general in steam locomotive
practice, had their origin in mid-western Europe. Egide
Walschaerts was one of the three Belgian giants. (The others
were Alfred Belpaire and, relatively much more recently, J. B.
Flamme.) Walschaerts' first patent was one of 1844, but more
important was his improvement of 1848. Quite independently,
in Germany, Edmund Heusinger produced in the following
year the radial gear on p. 139. The use of a return crank instead
of an excentric was Heusinger's work, and the same sort of
gear, in after years, was named for both men, according to the

(Centre) *Stephenson's link motion.*

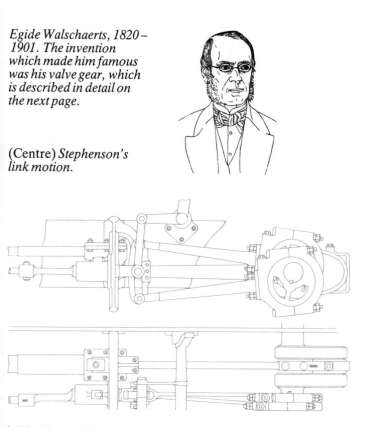

habit of people in this place or that. Neither reaped any patent royalties, but neither fell, like poor Richard Trevithick, into seedy poverty. Several much more recent American valve-gears have been of the Walschaerts/Heusinger family. An English gear of some interest, widely used in the late nineteenth century, was that of David Joy. From our drawing of the gear it will be seen there was neither excentric nor return-crank actuation. But this simplification was at the expense of perforating the connecting rod, which was all very well while locomotives remained fairly small, but apt to lead to bad breakages on big engines at high speed. It had quite a vogue in England on certain lines, notably the North Eastern and the London and North Western Railways. The Western Railway of France used it in a very British-looking 4-4-0 design (Series 900). In the early 'eighties, it was being installed at

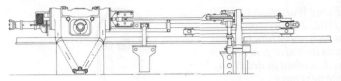

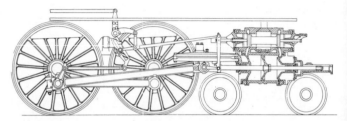

Walschaerts' valve gear was invented in 1844. It was quite different in arrangement from established link-motions and many English engineers,

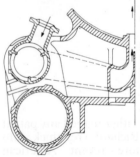

suspicious of such a radical change, were cautious of it. However, other countries adopted it and it turned out to be most satisfactory. It soon became the standard form on railways all over the world, including British Railways.

Altoona Works on the Pennsylvania Railroad, but already American locomotives were beginning to grow bigger than other people's, and making holes in their connecting rods was correspondingly less desirable.

At this time, slide valves were still practically universal, though ultimately, with the use of much higher pressures, the use of piston-valves became essential. As far back as the middle 'seventies, piston-valves were being made and installed on some London and South Western locomotives by William George Beattie, son of the man who was so clever with fireboxes and feedwater-heaters. But he was too eager.

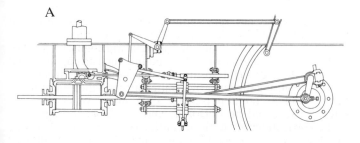

A *Heusinger's valve gear, 1849.*
B *Joy's valve gear.*

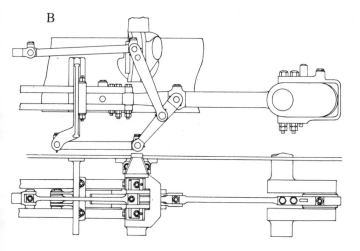

Metallurgical knowledge was still behind mechanical experience. There were some shocking failures, and as far as the South Western company was concerned, W. G. Beattie was "out", to be succeeded by W. Adams, sometime Italian Naval officer. Advanced theories can be expensive to a man. Fortunately Joseph Beattie had left his infant-prodigy well-off.

Boiler pressures rose, cautiously-gradual through the half-century 1850-1900. By the end of the century, 160 lb./sq. in., once prodigious, was common; 200 lb., later to be common, was prodigious.

Pressures suggest safety-valves. The old weight-loaded

safety-valve was not very suitable for locomotives, though it was used by both Kessler and Maffei in the German States for some years (e.g. in the Galician Carl-Ludwig engine on p. 134). Going over rough track or uneven points, such an engine would bump up the big weight, losing a lot of useful steam through the valve. Weighting the lever by a Salter spring-balance was much better, and a common practice over many years. But the safety-valve could be *doctored*, and accidents through such wicked operations led to the invention of other forms. Naylor's side-lever type, shown on Finland engine No. 11, still allowed interference, and also had a habit of sticking and then going off with a roar that shortened one's life.

One of the very earliest incorruptible safety valves was that patented by John Ramsbottom of the London and North Western Railway in the late eighteen-fifties, exemplified in the London and South Western engine No. 471. Its *Modus operandi*, like those of the valve-gears, is obvious, and it was much used everywhere but in North America, which in the eighteen-nineties speedily adopted direct-loaded safety-valves, in columns or cowlings which defied interfering enginemen seeking a little extra pressure. An admirable example was the Ashton safety-valve. Others included the Lethuillier-Pinel valve in France and the Richardson valve, much used in Northern Europe, especially Sweden and Finland. Neat little direct-loaded valves, locked-up in brass columns, were for long very popular in Scotland, on England's South Western line, and in Queensland, Australia, usually mounted on the dome. It was almost impossible to monkey about with any of these, or with the Wilson valve, once much favoured in Belgium, which superficially resembled the Ramsbottom valve but had

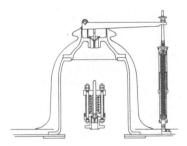

Safety valve.

the springs inside the columns, rising first against compression instead of by tension, and then by tension as it lifted conical caps on those columns.

Pressure was at first measured by a scale on the Salter spring-balance on the safety-valve. Bourdon's pressure gauge, which was to withstand the test of a century, had its index figure moved on a dial by the straightening tendency, under direct internal steam pressure, of a curved, flattened tube.

Hear the train blow! ran an old American ballad. There has been much argument as to the origin of the steam whistle, which later was to work just as well under air pressure on electric trains. The first steam whistle was invented by Adrian Stephens, a Cornishman, who went to South Wales like Trevithick, and possibly as early as 1826, applied such a whistle on a stationary boiler to the safety-valve, which also was equipped with a float against falling water level. It was certainly in use by 1833. The noise was produced by the impinging of an annular steam jet on an inverted brass bell or cup. In 1835, it seems to have turned up as a warning device for locomotives, and in the same year, quite independently, an English locomotive on the old Leicester and Swannington Railway was given a "steam trumpet" after an accident at a road crossing. The whistle was better than the trumpet, and since then has taken many forms and emitted many sorts of sound, from the joyous scream of England and Central Europe to the melancholy *whoo-hoo* chord of North America and Russia; from the frightened-lady squeals of French express engines to the deep majestic hoot—achieved by a miniature organ-pipe— remembered by many Scotsmen. French freight engines also sounded a deep note. There were such variations all over

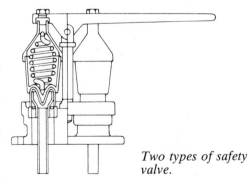

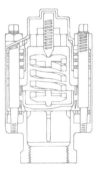

Two types of safety valve.

the world, according to influence by makers.

At first, locomotive boilers were fed by force-pumps worked off the engine proper, which meant that she had to move about to keep up the water-level in her boiler; sometimes an awkward business, if the subject were bottled up in some terminal station. The injector, whereby the operation was carried out by a jet of live steam from the boiler itself, was the invention of Henri Giffard, that most illustrious Frenchman who also produced the first navigable power-driven aircraft as early as 1852 (a steam-driven, hydrogen-filled airship). The principle was that the steam first exhausted all air from the appliance, then naturally achieved suction of the cold feed-water. The steam condensed and the resultant mixture, passing between opposed cones, had sufficient force to pass through a valve (the clack) into the boiler against the latter's pressure. The Giffard apparatus, first applied to locomotives in 1859, was the prototype of many later forms, though the earlier ones would work with cold water only. Hot-water injectors are relatively modern. Locomotives with early feed-water heaters, like Beattie's in the South of England, mounted steam donkey-pumps instead.

A prominent feature of locomotives in lands of wood fuel or difficult coal was the spark arrester. The old American "balloon" and "diamond" stacks have been already mentioned. An internal cone, inverted, and plenty of wire mesh, formed their business parts. An ingenious Scandinavian arrangement comprised a sort of fixed spiral turbine in a big collar at the base of the stack with the blast-pipe orifice above it. All were effective, but rather at the expense of spoiling the draught, and there was not much fast running with them. Later, less stifling, arrangements were inside the smokebox.

THE FORNEY

In the United States, the "American Type" (4-4-0, with outside cylinders) long maintained its position and was expanded by further coupled axles. The six-coupled version already was of some antiquity. Now eight-, and even ten-coupled versions appeared. On the Central Pacific Railroad, partner in the great transcontinental route of 1869 and later a constituent of the mighty Southern Pacific, A. J. Stevens' 4-8-0 locomotives climbed up and braked down both sides of the high Sierra in the early 'eighties. There was one 4-10-0, magnificently named *El Gobernador*, but a little before

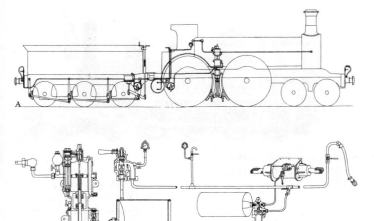

The Westinghouse Automatic Brake.
Fig A shows the Brake complete on an
Engine and Tender. Fig B is a diagram
showing the operation of the Brake,
from the Air Pump to the end of a
vehicle.

her time. Already, as we have seen, there were big and powerful locomotives in the mining valleys of the East, and of these we show the *Champion* of the Lehigh Valley Railroad in Pennsylvania, a very elegant example of the 4-8-0 type, though elegance was not what people generally expected in the Eastern coalfields of the United States, however beautiful were some parts of the country.

The 4-8-0 type was generally called "Mastodon" in the States after an extinct species of elephant whose giant fossil remains had lately turned up in the Tertiary strata of both Europe and America. Americans loved such type-names; we shall encounter others. The "Mastodon", however, was less widely used than the "Mogul" (2-6-0) and its 2-8-0 development for heavy freight, called in America "Consolidation". Both these were used all over the world. The "Mogul" on p. 103 was built for the Baltimore and Ohio Railroad in 1875.

A very different type of American locomotive was that first schemed by, and named after, Matthias Forney. It was in essence the common four-wheel American switching engine of

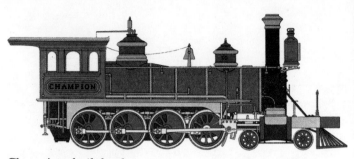

Champion, *built for the Lehigh Valley Railroad in 1882.*

Forney tank engine built in 1876 for suburban traffic on the New York and Harlem Railroad.

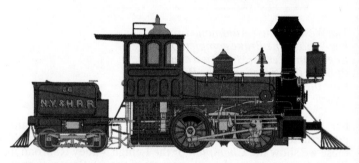

the mid-nineteenth century, but with the frames extended to take tank and coal-space instead of having a separate tender. It originated in the 'sixties, the first example having a bottle-shaped vertical boiler, a form usually deficient in steaming power. In the 'seventies, 'eighties and 'nineties great numbers of "Forneys" were built, sometimes for the rural, often narrow-gauge, "short lines" of New England and elsewhere, but most importantly for the city lines of which the New York elevated railways, trestled over the long-suffering streets, formed the finest examples. Though junior to the underground system in London, they made one of the earliest examples of *Rapid Transit* in cities, and were worked by hundreds of Forney locomotives until their electrification early in the twentieth century. For a "Forney" of the 'eighties, we show one built for the New York and Harlem Railroad.

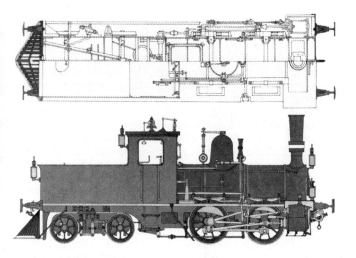

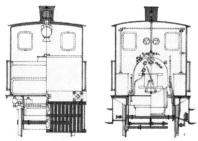

Forney locomotive built for the Thylands Railroad. Denmark. The designer was Otto Busse.

Derivation from the ordinary switching engine is particularly apparent in this, which looks more like that than a proper tank engine. Later examples were more compact, and larger, as far as weight restrictions on the trestle structures would allow. The New York and Harlem engine was rather a small suburban locomotive than an "Elevated" type, but she admirably shows the classic Forney arrangement.

These American city lines had their stations very close together, well in sight of one another down New York's long, straight Avenues. Driving an "L" train under steam was a matter of continually releasing brakes, giving steam, hooking-up the gear, shutting-off steam and braking to a stand all the way from Yonkers to down-town Manhattan, and back again. Lovett Eames' vacuum brakes were used, and this brings us to the subject of brakes.

America was the native country of powered brakes for trains, and to George Westinghouse, who first successfully applied braking by compressed air, on an experimental Pennsylvania Railroad passenger train in 1869, the whole world owes a debt not yet repaid. In its perfected form, the Westinghouse brake system depended on a powerful steam air-pump on the locomotive, constantly maintaining pressure in air reservoirs all down the train, supplying the cylinders which applied the car brake shoes, and it was *automatic in action*. That meant that in any case of failure (short of *losing the air* through arrant carelessness) all the brakes went on immediately. If a coupling broke, the connecting pipes broke with it, and the detached portion of the train came rapidly to a stand instead of colliding with the front part when that was pulled up by its anxious engineer; or, with even more frightful consequences, running backwards downhill and hitting something in rear. It may be remarked here that there was a very dreadful collision of this last kind, killing eighty persons, on the Great Northern Railway of Ireland in 1889. It moved the British Government of the time to make immediately compulsory the use of automatic power brakes on all passenger trains in the British Isles.

The culprit in this Irish case was a non-automatic vacuum brake invented by one S. Y. Smith (a compatriot of the great Westinghouse, one regrets to remark!). Air brakes would not do, however, on the New York Elevated lines; the short hops from station to station allowed no time for the engineer to work up his air pressure; hence the use of the Eames vacuum brake. Vacuum automatic brakes came to be used very widely in the British Isles and in various other countries, thanks largely to John Aspinall in England. The Tunisian Railways used a vacuum brake (Clayton's) until the early 1950s. Air was exhausted by a steam ejector, just as, in the Westinghouse and in later, kindred brakes, motive air was compressed by a pump. In the vacuum brake, application was by atmospheric pressure, always there, on the vacuum being destroyed. The appliance was thus not even dependent on a pump for its application; any mechanical failure meant that the brakes went on, or could not be released. Vacuum was less powerful than straight air, but sufficiently strong for the lighter trains east of the Atlantic, as for the very light, though often crowded trains of the American elevated and other city lines.

British horror of monopolies, comparable in the last century to American horror of socialism in the present, had absurd results, for some British railway companies used air, and some used vacuum. Through coaches and sleepers therefore needed to be dual-fitted (i.e. with two sets of brakes) to suit their various operating companies. There were two changes in braking between London and Aberdeen via the East Coast; one between Brighton and Plymouth in the South. Nor was Continental Europe faultless. The old-time Orient Express was air-braked twice and vacuum-braked twice on its long march from Paris to the Bosphorus (about 1,750 miles in sixty-five hours). Until the end of last century, its most important extra-Continental connection, a short one between London and Dover, was in the hands of two rival companies with different routes. Of these, the South Eastern Railway used vacuum brakes and the London Chatham and Dover Railway used Westinghouse air. No wonder some visiting Frenchman shrugged off the madness of the English as he stood surveying at Dover the two London expresses whose brakes, tickets and city terminals were all different!

Back to Matthias Forney's locomotive: Europe used it little. There was an isolated class on the Caledonian Railway in Scotland. In England the Yorkshire Engine Company built it for South Russia of all places. It appeared on a few Swedish and Danish branch lines. All these European examples, unlike the Americans, had plate frames, though the outline remained more-or-less American.

British railway companies, however, made much use of a

This little tank engine was designed by Dugald Drummond in the 1880s for the Caledonian Railways.

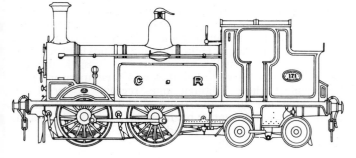

somewhat similar tank engine, still 0-4-4, but relatively compact, with inside cylinders, stemming from Stroudley's express engine which we have just noticed, and with capacious side tanks. Dugald Drummond, a fierce but gifted Scotsman who had been this Stroudley's Works Manager, was responsible for some of the most successful examples, though Samuel Johnson, who had been with both in Scotland and personally disliked them quite intensely, was first with the type on the English Great Eastern Railway.

We exemplify here one of Drummond's earlier specimens, built for the Caledonian Railway in the eighteen-eighties, for branch passenger service. The same designer, from 1897 onwards, built a very much larger version on the London and South Western Railway, capable of taking an express passenger train at a pinch, running both fast and steadily so long as the roads were decently solid. The little Scottish engines did not vanish until the nineteen-thirties, and of the sturdy London and South Western class we saw a last-survivor at Salisbury, in the South of England, as recently as April, 1964.

This type was almost unknown outside Great Britain, where it was widely and most happily used in Scotland, the South, the Midlands and the North East. Long ago, an American fellow-student, contemplating one and never even having seen a "Forney", complained that it was *all pfui*; to his eyes, back-to-front. It was nevertheless a most useful and remunerative type of locomotive; handsome too, in its portly way. It was many people's best investment.

The American Forney engine's effective survival was much shorter. It could not run fast, even when provided with an extra carrying axle in front, with a Bissell radial truck, as on the Illinois Central Railroad's suburban services around Chicago, making it 2-4-4. The 2-4-4 type of tank engine, however, was very successfully built in Germany by Krauss of Munich, with the Krauss-Helmholtz bogie. In this last, the leading coupled wheels shared a bogie frame with the leading carrying wheels, with suitable side-play in the coupling rods, only the driving axle being rigid to the main frames, for the remaining axles were on an Adams bogie. Helmholtz and Adams between them (they had their copyists) produced the two classic locomotive bogie trucks. The Bissell radial truck was by comparison a poor thing against these, though one regrets to record its considerable use in both America and the British Isles, sometimes with unfortunate results at high speed. The trouble with radials was that they suited only their own radii, while railway radii are variable.

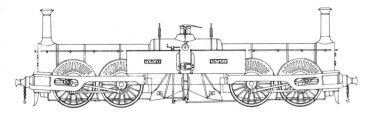

Articulated locomotive intended by Fairlie to be a fast passenger engine in France. There is no evidence that it was ever built.

THE FAIRLIE AND THE VICTOR EMANUEL

This business of flexibility in long and relatively heavy locomotives engaged many engineers, as we have seen in connection with the Semmering Trials of Austria. Robert Fairlie in England cribbed every idea he could and like many improvers got remunerative results. (In what Americans sometimes call "British English" he was, in some ways, *a nasty piece of work*.) The "Single-Fairlie" locomotive, like a "Forney" but with the motor unit also pivoted, with all the complications of flexible steam-pipes, was a wretched thing. New Zealand alone, on 3 ft. 6 in. gauge, managed to employ it with some usefulness on main-line service of a sort. (One of the engines once over-ran the pier at Lyttelton; its motor bogie dropped off into the harbour!)

Cesare Frescot built the first Vittorio Emanuele *type for the upper Italian railway.*

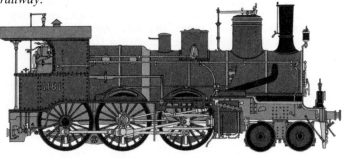

Much more interesting and successful was the "Double Fairlie", with two motor bogies and a double-barrelled boiler as in Cockerill's engine at the Semmering Trials. Several were built for minor railways in Wales during the 'sixties, and very small examples may be seen to this day there on the Festiniog Railway, now run by a preservation society. From the 'seventies onwards, large numbers of these Fairlie engines were built, some of considerable size, for Russia and Sweden, and most notably for Central and South America where gradients were very steep and curves very severe. We show an early specimen built for the Ichique Railway. The last and largest were for the Mexican Railway, well into the present century. Weak points were in the steampipes, which had to be either flexible or to have flexible joints, but these were largely overcome to produce a very powerful, very flexible, but rather slow locomotive. As the first two things were most important on steep mountain lines, and the defect scarcely mattered, the engines had quite a vogue in some places. One of the big Mexicans once ran away, from summit to foot of a most vertiginous mountain descent, reaching the bottom at far higher speed than her own steam might have managed, but undamaged with all wheels on the rails.

As suggested, the "Double Fairlie" was scarcely suitable for fast passenger trains, but that Fairlie himself optimistically planned such use is shown by a drawing signed by him and with annotations in both French and English, with metric dimensions. It undoubtedly shows a Fairlie express engine, though we have no evidence that it ever was built. The shape of the frames below the outside cylinders is that of Allan and of Buddicom in France, and whether desire outran achievement or no, one suspects that it was intended for the French Western Railway, whereon English influence was very strong.

As to the other important articulated type of steam locomotive in the nineteenth century, the Mallet, that must be deferred for a few pages as its history is tied up with compound expansion. Of orthodox types, the most important to appear from European works was the *Tenwheeler* (the 4-6-0, or 2-C) which America had been building for a long time as we know, though generally subject to light axle loads. Honours for European introduction might have been shared by Italy and Scotland (we are considering standard-gauge or broad-gauge engines for main-line traffic; there had been 4-6-0 tank engines for some time already). As Italy built the type for home traffic while Scotland built it at first for overseas use, let Italy have the kudos!

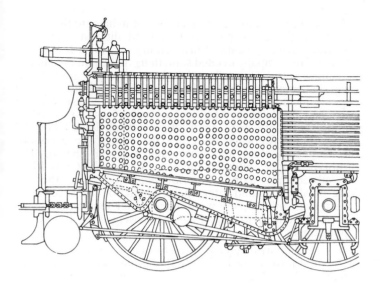

In order to burn low-grade coal, Belpaire built a greatly increased grate area. The old-fashioned iron grate was replaced by thin steel plates, rivetted together so as to allow air to pass easily through the centre surface. The old system of staying the fire-box was abandoned and replaced with horizontal and vertical stays. His first fireboxes had round tops externally but from 1864 they had the characteristic flat crown and flat-topped casing.

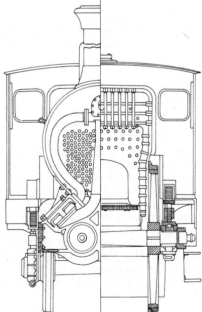

Cesare Frescot was Chief Mechanical Engineer to the Upper Italian Railways, and of its successor the Mediterranean System at Turin for many years of last century and for a few of the present. His company needed something that could take heavy passenger trains, as well as freight, on steeply graded and severely curved lines of Northern and Western Italy. (Special engines, including heavy tank engines coupled back-to-back in pairs, had long been used for banking on the Giovi Pass north of Genoa, but there were many other difficult lines.) For the Upper Italian Railways, therefore, its Turin Works produced the very competent and workmanlike design shown. The prototype engine, built for the new (*Succursale*) Giovi Line, was subsequently named *Vittorio-Emanuele II*, after the first King of all Italy, and later engines built for the succeeding Mediterranean company were named after eminent Italians. *Cavour*, needless to say, followed closely after the Royal Opportunist.

In style and arrangement, Frescot's earlier engines were an interesting mixture of French and German classical practice. There was little of British or American influence in them. The *Vittorio Emanuele* was, however, an entirely native Italian design. The drawing is sufficiently explanatory (there were two cylinders, with simple expansion) save that the engine was generally more massive than the American ten-wheeler of the period. The class continued to be built with slight alteration into the late 'nineties. It was not a graceful engine. The very short bogie in advance of the "works" prevented that. But it had a smart, soldierly appearance, with some of the brassy glitter soldiers were still expected to bear without being killed. From youthful days we recall sitting up and looking round, the first time we saw one. (It was in the engine's extreme eld, at Florence in 1922.)

Many "Victor Emanuels" were built and were found very useful over many years. In the year of the first, 1884, Dübs and Company of Glasgow (Scots firm with German founder) brought out a very handsome 4-6-0 engine for general service on the broad (5 ft. 6 in.) gauge Indian State Railways, up in the mountainous North-West of what is now Pakistan, though later examples were built also for India proper: for the Bengal Nagpur Railway and the Guaranteed State Railway (lovely title!) of His Exalted Highness the Nizam of Hyderabad. Immense numbers were built. They were still usefully employed in the nineteen-thirties, right up in the barren dun-coloured mountains of Scinde. An enlarged version by David Jones (who was probably the original, anonymous

Alfred Belpaire, 1820–1893.

designer for Dübs) appeared on the Highland Railway in Scotland, in 1894. In the same year, something outwardly like the Italian engines, though much enlarged, with four cylinders and compound expansion, came to the Gotthard Railway in Switzerland. The Swiss engines perished by electrification in the nineteen-twenties, but one of the Scots, now lovingly preserved, took the last steam train out of Inverness in the late summer of 1965. These were very handsome engines with big inclined outside cylinders centred over the long-framed bogie.

A quite different type of main-line engine, which in America was but barely represented, on the Philadelphia and Reading, and on the Chicago, Burlington and Quincy Railroad, was the 2-4-2 for fast passenger traffic. It was much used on the Orleans, State, and Paris-Lyons-Mediterranean lines in France—with outside cylinders as in the rare American examples—and owed its origin to Robert Sinclair, a Scot, who designed it for the Great Central of Belgium as far back as 1860 and was copied by the Moscow-Warsaw Railway a little later. The trailing axle greatly steadied, in the French examples, what was basically the vertiginous Stephenson "long-boiler" type.

Belpaire *2-4-2 express engine for the Belgian State Railways.*

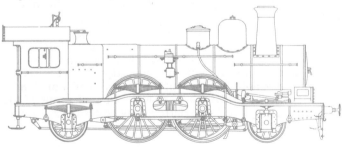

THE BELPAIRE FIREBOX

That great man Alfred Belpaire, however, produced quite a different form for the Belgian State Railways, which was built by Cockerill and others through the 'eighties into the 'nineties of last century. It had the very long, flat-topped firebox always associated with his name, capable of getting steam out of the most appalling muck the Railways Department could buy cheaply from the collieries of Charleroi. The cylinders were far forward and inside a set of very substantial double frames. It was an engine built like a battleship, of immense solidity. The driving wheels were large, but with plenty of steam behind them. The stack, very carefully calculated to promote a good even blast on the frightful fuel, was square in plan and tapered toward the top, in earlier examples, though later ones had the form of a truncated cone, likewise wider at the bottom. The design was built by Cockerill also for the Hessische Ludwigsbahn, with radial connection to the tender, but Belgium alone could stomach that square stack! A 2-6-0 version was built, with smaller wheels, for passenger traffic through the Ardennes.

Improvements in steaming efficiency, as well as mechanical and dynamic improvements, as in valve-gears and front-end passages, were becoming ever more necessary and important. Enormous freights were being handled all over the world, as well as coal and iron-ore. The feeding of great cities such as New York depended on wheat and stock trains rolling east from the Prairie States. More wheat trains in Canada, and stock trains in Argentina, fed London, thousands of miles across the Atlantic sea-routes. The Canadian Pacific Railway, backed by the legendary Hudson Bay Company, had reached the Pacific Coast in 1886; most South American railways were British-owned and equipped. In Northern Europe the great ore trains were beginning to rumble down to the Baltic from Lapland as well as from the Bergslag in Central Sweden. *Ergo*, bigger, more powerful and more efficient locomotives! People were travelling as never before; hence faster locomotives!

Coal was now the usual fuel, whether it were American anthracite, or best hard Welsh, or soft stuff from Belgium. After many experiments with water-partitioned fireboxes and combustion chambers, efficient combustion was very simply promoted by a deflector plate at the firedoor and a brick arch across the back of the firebox, though different forms of the latter suited different fuels; narrow long grates for good hard coal, wide ones for soft coal or low-grade anthracite, and so forth.

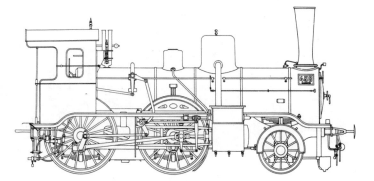

COMPOUND LOCOMOTIVES

Compound expansion, using steam first in high-pressure and then in low-pressure cylinders before exhausting, was one way to fuel economy, though complex. As far back as 1852, John Nicholson had experimented in England with what he called "continuous expansion", meeting with the frustration of many pioneers. Anatole Mallet, a Swiss, built the first entirely practical compound locomotive for the little Bayonne and Biarritz Railway, France, in 1876. Several systems of compounding appeared later in the century. Von Borries in Germany used two cylinders; a small-diameter one for high pressure steam and a large for low pressure, resulting in a curiously slow exhaust beat. A small von Borries compound passenger engine of the Hannover Lines, Prussian State Railways, is shown. The system was once common in Central Europe, and a form of it was still to be seen in Northern Ireland in the nineteen-forties.

Alfred de Glehn was an international figure, born in England of a Baltic father and a Scots mother, and by adoption an Alsatian. His system of compounding involved four cylinders;

Van Borries Compound locomotive, used for pulling passenger trains on the Hannover line of the Prussian State Railways.

(Right) *Alfred-George de Glehn, 1848–1936.*

155

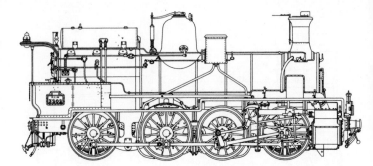

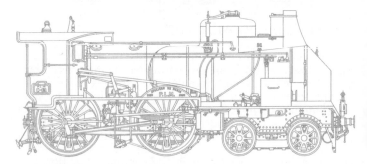

(Above) *Henry's 0-8-0 four cylinder
Compound Freight Locomotive,*
(Centre) *Henry's 4-4-0 fast passenger
engine, 1899.*
(Below) *Class C 4-4-0 of the Paris,
Lyons and Mediterranean
Railway, 1894.*

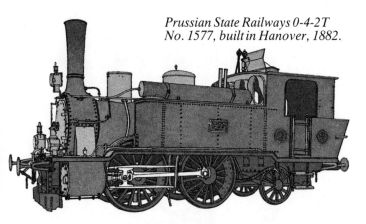

*Prussian State Railways 0-4-2T
No. 1577, built in Hanover, 1882.*

two for high pressure and two low-pressure, beginning with an engine built for the Northern Railway of France in 1886 (preserved today). His finest work must be ascribed to his partnership with Gaston du Bousquet of the latter company. In the 'nineties, successively bigger de Glehn compound engines came out on the Northern Railway of France and were paralleled on other lines. That shown was one of a set built in the late 'nineties for the English mail and passenger trains between Paris and Calais. One reason for the success of the de Glehn-du Bousquet arrangement was their scientific study and application of his theories on free steam passages, both admissive and for exhaust. The technics of these things need a complete book to themselves and here we must stick to general arrangement.

*Three-cylinder compound
coal engine, designed by
Francis Webb, 1893.*

The interior of a roundhouse on the Pennsylvania Railroad in the 1880s.

Both the French de Glehn-du Bousquet and the German von Borries engine, it will be seen, had their high-pressure cylinders outside and driving the rear axle, a disposition of T. R. Crampton's, with, except in the initial French engine, coupling rods to the preceding axle which was driven by the inside low-pressure cylinders.

We show next a heavy compound freight locomotive designed by M. Henry of the Paris, Lyons and Mediterranean Railway and built in 1888. In this the high-pressure cylinders were inside, between the leading and second coupled axle and driving on to the third. Many larger engines of the same type were built for the P.L.M. from 1893 onwards and were to be seen for over half a century longer.

Francis Webb, an English designer, at this time attracted much attention by his three-cylinder compound locomotives on the London and North Western Railway. He used outside high-pressure cylinders and a single very large low-pressure cylinder between the frames, with divided drive and, in the case of the passenger engines, uncoupled driving axles as in de Glehn's French engine of 1886. We show, however, a Webb three-cylinder compound freight locomotive, 0-8-0 like the Frenchman, built for the London and North Western company in and after 1893 for heavy coal traffic on this great British line. Webb's compound engines, though ingenious, had certain undoubted defects, though hundreds were built. They *could* do excellent work, but it needed an artist to drive one. Maintenance was heavy. Austria, France, India and the Americas sampled the type–but no more!

Of American compound locomotives, those under Vauclain's patents mounted high- and low-pressure cylinders together, outside, with a single cross-head to each pair of piston rods. Many such did well, but the arrangement entailed very heavy reciprocating masses. The same defect applied to various compound locomotives in both America and Europe which had the high- and low-pressure cylinders arranged in tandem with common piston-rods.

Karl Gölsdorf in Austria produced some remarkable compound locomotives under his own patents, which hauled both passengers and freight over the Alpine passes—Arlberg, Semmering, Tauern and Brenner. They were variable engines; earlier examples steamed badly, and an old Gölsdorf was not to be compared with the de Glehn-du Bousquet engines, working very fast express trains on the great French main lines. At one time and another, the finest fast express work in the world was to be found between Paris and Calais.

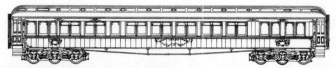

(Above) *Pullman car of the type in common use in America, c. 1898.*
(Centre) *In 1894, the North Eastern Railway built this sleeper, designed by David Bain, which contained the first single-berth compartments in Great Britain.*

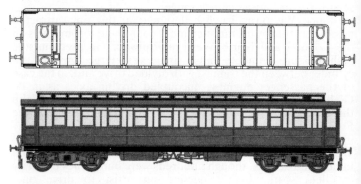

SLEEPING CARS

Mention of the increasing weight and speed of trains turns us to car design (*carriages and wagons* in all British or British-influenced countries). In passenger service, we have seen how American design was improved by G. M. Pullman. By the end of the century, the native Pullman car was an enormous thing; still built chiefly of wood but spacious, of great length and formidable weight. We show a standard Pullman sleeper for United States service, built in 1898. The wide vestibules and closely engaging gangways with friction plate contact, one to another, were the work of Henry Sessions of Pullman's works, one of the truly historic coachbuilders. The clerestory was to remain supreme in American passenger car design for many years.

America clung faithfully, and for as long, to the old centre-aisle arrangement of curtained sleeping berths, though private compartments, reached by side corridors, could be reserved for a higher fee. European users invariably preferred

(Above) *The interior of an early
sleeping-car of the Pullman type.*
(Centre) *A dining car in Germany in
the 1880s.*

these and were prepared to pay for them, even though on older
cars it might entail sharing a small bedroom with three total
strangers, as in the Mann-Nagelmackers arrangement.

Viaduct over the Maderaner Gorge at Amsteg, Switzerland, 1880s.

To the best of our research, the first standard sleeping cars to provide single-berth cabins for ordinary passengers appeared on the East Coast expresses between London and Edinburgh, in 1895, the work of David Bain of the North Eastern Railway in England. These cars also provided double-berth compartments for couples, and we show the original outline. Apart from the wide vestibules (at first without connecting gangways), that outline suggests a much older American car. But—oh blessed single bedroom! Bain's car became a prototype for all Europe, where the Wagons-Lits company took it up later for the more luxurious international expresses.

Sleeping cars were generally for first-class passengers only, though Russia, the Western United States, and Canada (which delicately called the vehicles "colonist cars") provided rather Spartan folding berths for unmoneyed migrants. As to heating, while England had a touching affection for the portable metal foot-warmer on the carriage floor, stoves of variously frightful sorts abounded. America, which could be *very* cold in winter, used the Baker heater, a stove-fired, closed-circuit, hot-pipe system using either water or a saturated saline solution which would not freeze when out of use. On the Belgian Great Central Railway (1875-76) M. E. Belleroche furnished a highly advanced hot-water circulation system throughout the train, using tender water from the engine, warmed by a special injector. In spite of the hazards of early hose connections, it answered quite well. Contact with the passengers was through copper floor-plates called chaufferettes which cooked one's boots and anything else inadvertently dropped or spilt on them. Warm-water heating was general in the Netherlands.

Low-pressure steam heating of trains from a reducing valve on the locomotive ultimately superseded hot-water systems (introduced on the Eastern of France in 1874). W. S. Laycock's, with steam piped to storage heaters under the carriage seats, appeared in England in the early 'nineties, and other systems gradually followed, though in old English local trains the wretched travellers continued for some time to sit and shiver in winter.

Lighting greatly improved. England pioneered gas lighting of trains as far back as 1863. Later, compressed oil-gas superseded coal gas, as in the Pintsch system. Kerosene lamps were still much used in America, and candle-lamps in Russia. The first electrically-lit car was a solitary Pullman in England, on the London-Brighton line in 1881, but the necessary batteries were clumsy and very heavy. Until, right at the end of

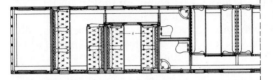

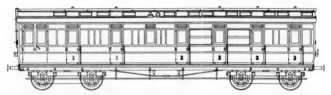

"Lavatory Bogie Composite Coach"
built in 1901 for the Great Eastern Railway, England.

the nineteenth century and in the British Isles, J. C. Stone perfected self-generating and self-regulating electric lighting equipment, oil lamps and gaslight held their own in most places. A great pioneer of electric light was the Great Northern Railway of Ireland which *never* used gas. Both gas and petroleum were liable to set trains ablaze after an accident.

Passenger communication was for long very bad. *Pulling the cord*, which *might* ring a bell, was the alarm signal in many places. Passenger's emergency-access to the continuous brakes was first provided on the Grazi-Tsaritsin Railway in Russia, by its Scots engineer, Thomas Urquhart, in the 'eighties. It was the precursor of all modern systems.

While the American ideal of aisled or gangwayed cars spread to much of the world, closed compartments remained usual for coaches in many Western European countries, and are well exemplified by this English carriage, one of a series built for the Great Eastern Railway at the turn of the century. Though small, its third-class compartments and its water-closet access were both excellent. As long ago as 1875, the mighty Midland Railway, stretching from Bristol and London to the North of England, had cushioned the seats of *all* coaches, a most radical reform which other British companies had to copy to compete with each other. In form this Great Eastern carriage was much akin to Prussian practice of the period, though not in its provision of third-class plush.

The best Prussian carriages of the 'nineties, the famed *D-Wagen*, were prototypes of the usual Continental European passenger car of today; vestibuled and gangwayed like

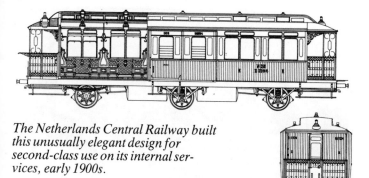

The Netherlands Central Railway built this unusually elegant design for second-class use on its internal services, early 1900s.

Pullman cars, but internally furnished with compartments and side-corridors. Ordinary French carriages of the same decade, alas, were a reproach, though the best cars could be handsome enough.

In Eastern Europe and Scandinavia, the American plan was widespread, though modified. In the Netherlands this was favoured only by J. W. Verloop on the Dutch Central Railway, of which we show an elegant second-class car for local traffic. (The rigid-wheelbase arrangement of six wheels, by the way, was unknown in North America. It was rough-riding, yet steady.)

Only in England (not even in Scotland!) and in North America at this time, did one find any sort of ornament about coaches for passengers paying the minimum ordinary fares. In Continental Europe, however, the *second* class was often elegant as well as comfortable, while in England, where the *third* class rapidly improved, second class became gradually moribund on the main lines. Two classes were enough, like day-coach and Pullman in the States.

Dining cars are of some antiquity. The first of which we have actual record was a convertible diner-sleeper (an "hotel car") on the Great Western Railway of Canada in 1867, about two years after Pullman's sleeper *Pioneer* in the States, though there was a diner in South Russia about the same time. A straight Pullman dining car appeared on the Chicago and Alton Railroad in 1868. The first dining car in Western Europe was a converted Pullman parlor-car called *Prince of Wales* between London and Leeds on the Great Northern Railway (England) in 1879. In all cases the term implies meals cooked in the

165

same car *en route*.

Both Old World potentates and New World business nabobs at this time travelled in private cars—and, in the case of the monarchs, in whole special trains—of great sumptuosity. Some of the richer Americans, indeed, outdid the kings during the late 'nineties and early nineteen-hundreds. One doubts that there was ever anything on wheels more sumptuously decorated than the State coach of Maximilian II and Ludwig II of Bavaria, but their train had no bathroom. Nor had Queen Victoria's in England. But Colonel William Jackson Palmer had one on the Rio Grande Western in 1892, and so had other men like him. One reads even about marble tubs and gold-plated faucets.

Even those not quite in a position to own or permanently command such things were able to hire them in the United States. In Europe a few had them, like the Chancellor Prince Bismarck and the Duke of Sutherland. With the British aristocracy and richer bourgeoisie, the hired family saloon was a recognized institution. It was sufficient to pay so many fares and then temporarily set up house in one.

An artist's impression of the smoking lounge and library of an Orient Express in the 1880s. (Courtesy of the Radio Times Hulton Library.)

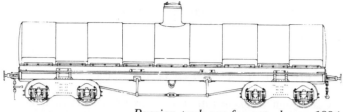

Russian tank-car for petroleum, 1894.

Mails had been carried by public transport in the days of the Caesars, though the service doubtless went down in the Middle Ages, when a mounted, armed courier carried them. Postal authorities quickly recognized the value of the rail. The Travelling Post Office, which meant not only the carriage but the reception and sorting of mail *en route*, originated in the English Midlands in January, 1838, at the instance of F. Karstadt, doubtless a German by descent. By May, John Ramsay of the British General Post Office had devised apparatus for picking up mailbags without stopping. Ten years later, John Dicker, also of the British G.P.O., had perfected apparatus for both pick-up and delivery of mail-pouches at full speed, still widely used on the great British main lines, and, at one time and another, in Prussia and France, though less extensively. The first French Travelling Post Offices worked between Paris and Rouen in 1844. Post Office sorting cars made their first American appearance on the Hannibal and St. Joseph Railroad during the Civil War, at the instance of W. A. Davis, an inventive postal clerk at "St. Jo." The *T.P.O.* spread.

In America it was the *R.P.O.*, the Railway Post Office. In an age of mail-bearing aircraft, the postal train holds its own on overnight inter-city service by its virtue of being an office as well as a conveyance. In pioneering England the all-postal trains to-and-from Scotland and the West run as of yore. They have no room for passengers.

Of European freight-cars in the late nineteenth century, one cannot say much, but the great distances of both Russia-Siberia and North America demanded—and got—vehicles of very substantial sort. We show here a tank car for petroleum, built in 1894 for the Russian line between Baku on the Caspian and Batoum on the Black Sea. The American freight-car of this time was a huge oblong box with a cat-walk along the top for the nameless heroes who, in the iciest weather, skipped from brake-wheel to brake-wheel as their train screamed down some vertiginous grade, were it in the Rockies or the Alleghenies.

From such horrors was George Westinghouse beginning to deliver them. The diamond-frame truck was its tower of strength.

THE FIRST ELECTRIC LOCOMOTIVES

Car design brings us rather abruptly to the very first electric trains, themselves closely akin to street-cars or trams of American origin. George Stephenson himself had foreseen the ultimate triumph of electric power. Germany was the cradle of electric traction, though there had been experiments in many places. Behind its advance was the great Siemens family, of whom Wilhelm Siemens became an English knight though it was Werner von Siemens who first successfully conveyed passengers, on a narrow-gauge line in the Berlin Trades Exhibition of 1879. Haulage was by a motor mounted over and geared to four solid iron wheels, taking current from a third rail. The motorman sat astride. On May 12, 1881, a tram-car with electric motors inaugurated the world's first public electric railway at Lichterfelde, Berlin. On August 3, 1883, Magnus Volk opened the first section of his famous narrow-gauge electric railway along the sea-front of Brighton in the South of England. It is still there; antique but well-patronized in summer, and affectionately regarded by many.

It was in Ireland that water power was first used to generate electric energy for traction. In the words of its sponsor and

At the Berlin Exhibition in 1879, this little electric locomotive pulled the illustrated "train". It was designed by Werner von Siemens.

builder, William Acheson Traill, the Giant's Causeway, Portrush and Bush Valley Railway and Tramway Company, on the 3 ft. gauge, "was a bold venture, a Utopian scheme", for he proposed "a new traction power to supersede the long established horse or steam power for the working of tramways or railways; and, wilder still, the idea of utilising the waste forces of our rivers to generate electricity to propel tram-cars along a tramway miles away from the source of power." Traill's consultant was Wilhelm Siemens. Turbines and dynamos were installed at a 24 ft. fall on the River Bush, though here there was some delay and spare steam-powered generators were installed at Portrush. A trial trip was made on November 21, 1882. Teething troubles were many, and two steam-tram engines were wisely acquired, but regular electric services began on November 5, 1883. Steam was used in the streets of Portrush, where the raised electric conductor rail was out of the question. Indeed for many years the little line was as much what the Hollanders called a *Stoomtram* as an electric railway. Overhead contact ultimately replaced the troublesome *hot rail*. Tramway or railway, it was the ancestress of mighty electric lines powered by falling water, stretching from Lapland to Sicily, and in many other places both European and Asiatic.

The world's first electric underground line was in England, where the first part of the City and South London Railway was opened by the Prince of Wales (later King Edward VII) on November 4, 1890, using very small electric locomotives to haul three-car trains through deep-level tunnels in the London

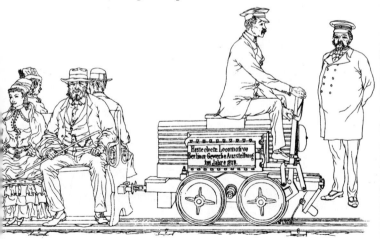

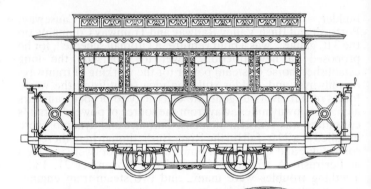

(Above) *In 1883, Ireland saw the first use of electric traction with hydro-electric power on the little Giant's Causeway line from Portrush. This is a 20-seat car built for the line.*

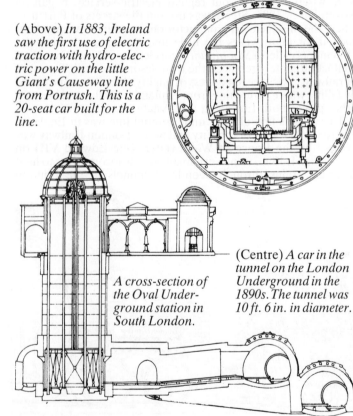

A cross-section of the Oval Underground station in South London.

(Centre) *A car in the tunnel on the London Underground in the 1890s. The tunnel was 10 ft. 6 in. in diameter.*

clay. Although a rather fearsome conveyance (the citizens at once named the narrow, almost windowless cars "padded cells") it was an immediate success. Electric underground lines were ultimately to serve most of the great cities of the world.

Though America had pioneered elevated city railways, it was England that first built one with electric traction, the Liverpool Overhead Railway of 1893. Elevated city lines were less favoured than underground ones. Today, they are gone from both New York and Liverpool, though they may yet return as traffic problems worsen, possibly in the form of monorails, the oldest of which, at Wuppertal in Germany, dates back to the end of last century.

But back in the last quarter of last century, the electric train was still a mechanical curiosity. In Scotland, the usually shrewd city of Glasgow so far distrusted it as to build an underground line with cable traction (1896) and to regret this error of mechanical judgement for the next thirty-nine years. (The line was electrified in 1935!)

SPEED RECORDS

The sun of steam was still rising to zenith. Except for use of superheated steam, which really belongs to the present century, advances meant increase in size, above all in North America. By the end of the century, American locomotives were powerful, effective, and in some cases both speedy and very handsome. In Continental Europe there was an ugly phase. British engines were elegant but usually small.

The *American Type* (4-4-0) remained the *old faithful* for fast passenger service, almost to the end of the century, and perhaps the finest examples of this were Theodore Ely's on the Pennsylvania Railroad, and William Buchanan's on the New

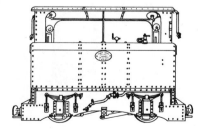

This electric locomotive was built for the City and South London Railway in 1890 by Mather and Platt in association with Beyer Peacock. It ran on the first tube railway in the world.

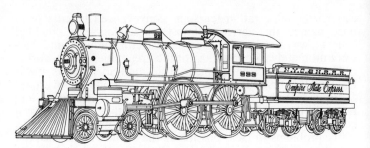

The New York Central No. 999, 1890s.

York Central, though other lines had many worthy designs. If an Anglo-Saxon author may criticise America, let him remark that Buchanan's achieved the greatest international fame, while Ely's were most beautiful to the eye.

New York Central No. 999 is (for she is treasured to this day) a Paul Revere among locomotives. By our records she was the first man-made machine to run at a land speed of 100 miles an hour or more, by responsible calculation. (There had been hundreds of irresponsible claims, almost since locomotion began!) She was built in 1893 and was a star turn at the Chicago World Fair of that year; furthermore, she was designed for a speed demonstration rather than for general service (later her driving wheels were reduced for that!). With the original 86 in. driving wheels, on May 10, 1893, No. 999 was timed at 112.5 miles an hour over a measured mile near Batavia, N.Y., with a four-car train. This was an official road test, though without a dynamometer car, and there is no reasonable cause to doubt its authenticity. It was to remain a world record for some years, though European opinion, jaundiced by many older and quite irresponsible American claims, was sceptical. The fastest authentic maximum speed previously

William Buchanan was born in Dumbarton, Scotland, in 1830 and emigrated to the United States. He designed the New York Central 999 (above).

172

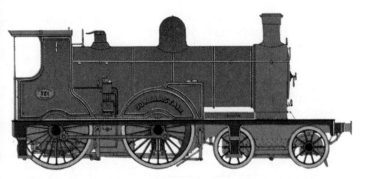

had been one of 89.5 miles an hour attained by an ancient Crampton locomotive, Eastern Railway No. 604, rebuilt with a patent boiler by Flaman, on the Paris-Laroche line (June 20, 1890).

Speed-for-speed's-sake was indeed something of a fetish at this time with competitive railroad companies, as between New York and Chicago in the United States, and London and Aberdeen in Great Britain. Over the latter stretch there was a regular race, night after night, during the summer of 1895,

(Above) *John F. McIntosh designed the* Dunalastair *to be a standard passenger locomotive for the Caledonian Railways, 1896. The type was also built for the Belgian State Railways.*

(Below) *The Swedish State Railways' E2 Class 2-8-0 inside-cylinder engine, 1908.*

using quite ordinary English and Scottish locomotives over rather exacting roads. Average speeds were what counted. The highest were attained by the West Coast Route (London-Carlisle-Perth-Aberdeen) on the night of August 22-23, when the 540 miles were covered in 512 minutes with three station stops of two minutes each. The fastest "hops" were at 67.5 miles an hour from Crewe to Carlisle (142 miles, London and North Western engine *Hardwicke*, Driver Robinson) and Perth to Aberdeen (89.7 miles in 80.5 minutes, or just under sixty-seven miles an hour with Driver Soutar on Caledonian Railway engine No. 17). Excellent work was done by both routes, allowing for the lightness of the trains (about four small wooden coaches or sleepers). The little engine *Hardwicke* is preserved in London. As far as any lessons were drawn from these picturesque pranks, they were, firstly, that locomotives would take bigger payloads if the proportion of boiler power to cylinder capacity were increased, and secondly—a negative one—that most passengers had little wish to be turned out into a still sleeping city with little prospect of breakfast. In the following year, the London and North Western company had a bad derailment at Preston through high speed round the curves there, and further "racing" was voted off.

But in that year, the standard Caledonian passenger locomotives acquired much bigger boilers, as in John F. McIntosh's *Dunalastair* which basically was of Dugald Drummond's old type dating back to the 'seventies. So successful was this modification that numerous "Dunalastairs" were built under licence for, of all outfits, the Belgian State Railways, which used them on the Brussels-Ostend and Brussels-Antwerp lines. The 4-4-0 engine with inside cylinders remained for years a British national type—in India and Australia too—as well as being much used in Holland and Sweden. In France it was rare, except on the Western Railway. In Central Europe it was almost unknown (exception in Baden) while in America there was no revival of the "inside-connected" engine. It made a strong and solid engine, but the machinery was somewhat inaccessible.

THE SINGLE-DRIVER REVIVED

In France, however, we see during the 'nineties some early attempts by modified external surfaces to reduce air resistance on fast passenger locomotives. Baudry's work on the Paris, Lyons and Mediterranean Railway could not be called streamlining; "windcutting" was the vernacular word (*coupe-vent*), and it involved wedge-shaped casings to smokebox, stack, and the front of the cab, and lenticular casings to the domes. F. U. Adams had tried something of the kind, and with fully streamlined cars at that, in the United States during the 'eighties, but the time was not yet ripe for these. Ricour tried the *coupe-vent* style on the French State Railway as far back as 1887, too. The odd thing was that the mighty P.L.M. was by no means "fast" by French standards. Its trains were often heavy; they rumbled rather than raced.

There was one feature of steam locomotive working peculiar to Great Britain, the French State Railways between Chartres and Bordeaux, and to certain lines in the Eastern United States, but unknown everywhere else. That was the laying of immensely long water troughs on level stretches of track, whence the engines took water at full speed (in good American "on the fly") so that very long runs could be made without need for water stops. It was an English invention, by John Ramsbottom of the London and North Western Railway, and dating back at least to 1860. Pickup was by a hinged scoop, on the tender. The arrangement lasted for as long as steam traction did; it accounts for the very small tenders of many otherwise large and important British locomotives down the years.

Invention of steam sanding gear caused, on British lines, an astonishing revival of the passenger locomotive with single driving wheels, which had never quite vanished from them. Such engines continued to be built to the end of the century, and we show a very handsome example by Samuel Johnson, the Midland Railway's *Princess of Wales*, built in 1900. Remote descent from the Stephenson "Patentee" is discernible. The engine was shown at the Paris Exposition of 1900, causing some French astonishment but winning a gold medal. Some of

The de Glehn/du Bousquet Compound Locomotive No. 2160, 1897.

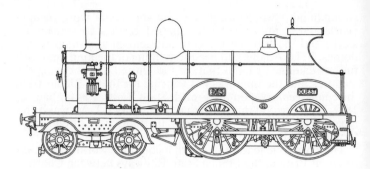

*A 4-4-0 express, showing English
influence on the Western Railways of
France at the turn of the century.*

these Midland "single-wheelers" lasted into the middle nineteen-twenties. They were extremely fleet engines and could move quite surprising loads.

Small single-drive tank locomotives were built for local services over many years, notably in Austria and Sweden; the last for Latvia in 1928, but outside England (which built four very large ones for China as late as 1910) express locomotives with single driving wheels were by now extremely scarce. There were the short-lived "Bicycles" on the Reading Company's Atlantic City service, also in the 'nineties, and a few (British-built) in Argentina. Those Reading "Bicycles"

*Princess of Wales, end of
the single-driver legend in England.*

176

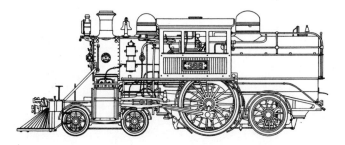

had a virtue purely incidental to their single driving wheels. Use of a carrying axle at the rear allowed for a much larger firebox of the sort pioneered on the same road by Millholland. American designers now went for a coupled passenger locomotive with this same advantage. Briefly the 2-4-2 type, already much liked in France, was given a trial, but the important development was the 4-4-2, known immediately as the Atlantic type from having made its first appearance on the Atlantic Coast Line in 1894, from Baldwin Locomotive Works. It had the steadiness the 2-4-2 type lacked, as well as a more adequate boiler. We show a similar engine, built soon after for the Concord and Montreal Railway linking New England and Canada. By 1896, American "Atlantics" were being built for the Central of New Jersey, Lehigh Valley, and other railroads, with very wide fireboxes and with the engineer's cab mounted saddle-fashion on top of the boiler. They were often called "Camelbacks", though

(Above) *The Reading "Bicycle" was the brainchild of L. B. Paxson and was designed by W. P. Henszey of the Baldwin Locomotive Works.*

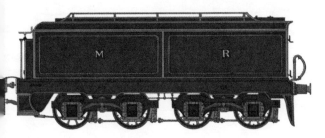

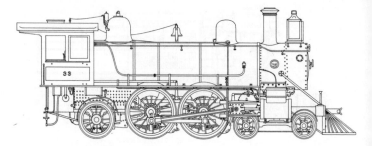

*This Atlantic Type locomotive did
heavy duty on the Concord and
Montreal Railroad for many years.*

this tended to confusion with the ancient types of Winans and
Hayes on the Baltimore and Ohio. "Mother Hubbard" was at
once a less confusing and more endearing nickname. The
unfortunate fireman had a wet and windy perch in rear of the
firebox.

In the same decade, the Atlantic type arrived in Europe.
There were 4-4-2 express engines with inside cylinders on the
Palatinate Railway in Germany, and on the Lancashire and
Yorkshire Railway in England, but the true Atlantic had the
cylinders outside, driving the second of the two coupled axles,
and this was built in England by Harry Ivatt of the Great
Northern Railway from 1898 onwards. The original engine,
soon after named *Henry Oakley*, is preserved at York. Still,
boilers were inadequate; much bigger ones could go on to such
a locomotive, and so they did in the early years of the following
century. Ivatt's later examples in England were to give useful
service through nearly half a century, including two major wars.

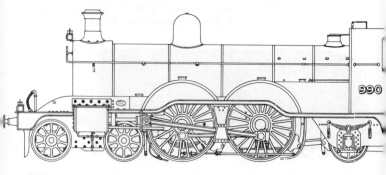

In America, the Pennsylvania Railroad was to produce the type to absolute maximum dimensions, and one example, No. 460, achieved a latter-day record when Colonel Charles Lindbergh made the first Atlantic solo flight in 1927 with the aeroplane *Spirit of St. Louis*. The films of Lindbergh's arrival were developed in a Pennsylvania baggage car for immediate showing on Broadway, and the engine in question covered 216 miles in 175 minutes. Be it remarked that this was, by then, with an old-type locomotive (P.R.R. Class E 6) taken straight from stock. The world's last "Atlantics" in steam were some of Denmark's beautiful Class P, still about in the nineteen-sixties.

Back to the eighteen-nineties; a funny thing happened in America. Just as certain Englishmen believed that only with a single driving axle could one get a "free-running" engine capable of really high speed, so did many Americans believe that no more than two coupled axles were permissible to the same end. Six-coupled wheels were all very well for fast freight, or for heavy passenger hauls over mountains, but—

At that time and for many years, there was intense competition for the New York-Chicago traffic by, respectively, the Water-Level Route of the New York Central and its ally the Lake Shore and Michigan Southern Railroad, and the rival route over the Alleghenies, which was a preserve of the Pennsylvania Railroad. The rival business interests, those of the Vanderbilts and the Depews, were more than just that; they were mortal enemies, but on the whole they confined their warfare to rates and rival services. (It was the rascally Jim Fisk of the Erie Railroad who got himself shot dead; and that was over a girl, though doubtless there were dry eyes in the Vanderbilt family on its being informed.)

But "Chicago in Twenty-four Hours" was the ideal of both parties in New York. By a chance of rostering, one of the

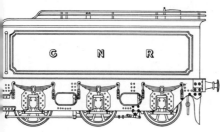

GNR No. 990, 4-4-2 express passenger engine, designed by Harry Ivatt for the Great Northern Railway of England.

Water Level Route's demonstration trains, on the Lake Shore and Michigan Southern Railroad, was headed by a little fast-freight 4-6-0 engine (No. 564) between Erie and Buffalo Creek, and covered the 86 mile stretch at a start-to-stop average speed of 72.91 miles an hour. To conventional American thought, it was as if someone had won the Kentucky Derby with a plough-horse. The "big wheel" idea died in that hour, as far as America was concerned. There were no more "Bicycles". Take a free-running engine, with good front-end arrangements, as in admission and exhaust, and a boiler making plenty of steam, and speed as well as power would look after itself! Already the Chicago, Milwaukee and St. Paul Railroad had produced a 4-6-2 type for heavy passenger service, though the extra axle had been by way of lessening axle loading and not, as in the Atlantic type, as a support for an extra-large firebox. It was not yet the Atlantic's successor, the Pacific Type locomotive. That would belong to the next century.

But the ten-wheeler–the 4-6-0 engine–which for so long had been the recognized American fast freight locomotive, came into its own for heavy passenger haulage while the 2-8-0 took more and more of the country's freight and mineral traffic. In the same decade, these two arrived in Russia also, and more and more in Western Europe.

In the Alpine lands, Anatole Mallet's semi-articulated locomotives found their numerous "feet" at this time. It was in the next century that they were to take North America by noisy storm. The essential feature of the Mallet locomotive lay not only in its two sets of coupled wheels, but in the fact that only the first set was pivoted, forming a motor bogie. The after set

The 0-10-0 freight locomotive was much favoured in Russia. This was one of 66 delivered by Nydqvist and Holm of Sweden.

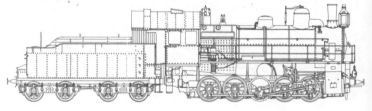

180

was mounted in rigid frames; hence our term "semi-articulated", for the fully articulated engine was exemplified in double-bogie designs, as of Fairlie and Meyer.

Further, Mallet had been primarily interested in compound expansion, and his original conception had been that of a compound locomotive mounting its low-pressure cylinders on the leading bogie, thus having all axles powered while the engine as a whole was flexible to an extent impossible with the multi-coupled types of Central Europe and North America. There was less difficulty with flexible steam-pipes when these involved only low-pressure admission. Mallet's effective experiments took place in the late 'eighties, as the Englishman Fairlie's had done in the 'sixties. In 1890, under his patents, Maffei of Munich built for the Gotthard Railway in Switzerland what was then the largest locomotive in Europe and one of the largest specimens in the world, even allowing for North American prodigies. She was a tank engine, double 0-6-0, or, *auf deutsch* C-C. Her fault was that common one of the time; her boiler was not big enough. Had it been so, her axle-loading, with the track standards of the day, would have been too high.

At first, smaller examples of the Mallet type were much more successful, notably on the metre-gauge Landquart-Davos Railway in Eastern Switzerland and in a design that was practically identical on the Yverdon and Ste. Croix Railway in Western Switzerland and on the Provence Railway—the lovely, vanished *Ligne du Sud de la France*—both also metre-gauge. A very fine early Mallet tank engine on standard gauge, once belonging to the Swiss Central Railway, is preserved in the Verkehrshaus at Lucerne, in all her splendour of green and red.

*J. A. Maffei's Mallet compound
locomotive for the Gotthard Railway,
Switzerland, 1890.*

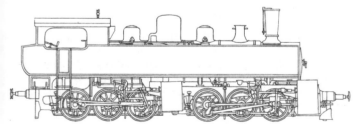

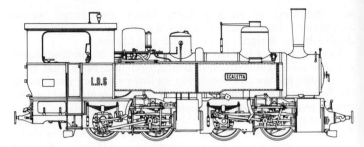

Scaletta, *one of the earliest Mallet
compound semi-articulated locomotives. Built by Maffei for the Landquart-Davos Railway.*

By the end of the century, the steam railroad had conquered
most of the world's great land masses, including some of the
greatest mountain ranges. The locomotive was a part of
people's lives as no machine had ever been before. It conveyed
the emperors and the kings and nabobs; it hauled the emigrants
from Ukraine and Silesia to the Channel Ports, and from
Montreal or Jersey City to the Far West. It took wealthy
tourists to classic lands and cheap trippers to the mountains or
the seaside. It took artisans and clerks by the million in and out
of the world's greater cities. As for freight, the steam train had
become the overland ship. But already new challengers had
appeared over the horizon.

We have noticed already the modest arrival of the electric
train, at first as little more than a mechanical side-show. But
when people had seen the first electric underground city
railways, steam was doomed below the streets. Certainly, the
steam underground lines of London—and, from the
middle-eighties onwards, of Glasgow and Liverpool—were
cordially detested. Their atmosphere at busy times beggared
description, like something out of Dante or Milton.

It was Baltimore, in the United States, that saw the first
transformation. There, between the Waverly and Camden
stations, was a succession of excessively foul tunnels, and this
was included in a three-mile stretch electrified by the great
Westinghouse company in 1895. Haulage was by paired or
articulated locomotives, each double unit on four fully-
motored wheels to make what we would nowadays call a Bo +
Bo machine, collecting current from an overhead *rail* (not a

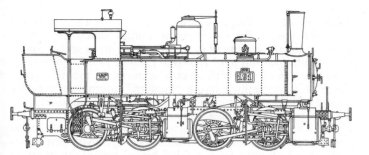

J. A. Maffei's Mallet tank engine,
Class C4, No. 191 on the Swiss Central
Railway, designed in 1890 and built in
1894.

wire supported by catenaries, such as we know) by means of a collector shoe on a folding steel arm suggestive of a draughtsman's pantograph. These "motors", as they were at first called, took entire trains in and out of Baltimore, the steam train-engines being hauled together with their cars, with steam off and valve-gear neutral. This was the first movement of full-size trains by electric power on a standard-gauge railway. It was an undoubted "American first", to the particular credit of the Baltimore and Ohio Railroad which, years before, had taken the first train into Washington. One could see the shape of things to come through the mirk of other city railways, and also of the great mountain tunnels.

But there were other shapes of things to come, which common observers took even less seriously than they took electric traction. On the roads of both Europe and America, the motor-car had emerged from the experiments of Karl Benz and Gottlieb Daimler. From 1896 onwards, even cautious England allowed it to proceed without having a red-flag man in front, while France and America received it with delight. Not yet did people imagine the motor-car as a rival to the train itself. Its capacity was small; it was extraordinarily expensive; it was a sport of the rich.

Internal combustion locomotives are younger than motor cars on the road, which is odd, seeing that Karl Benz was initially a locomotive engineer. Not so odd, perhaps, considering that revolutionaries are often entirely so!

Briefly to recapitulate mechanical history: The first practical internal combustion engine was of course the gun, though it did not make anything *go* except some poor wretches' arms and legs. Huygens in the seventeenth century aimed, though no more, at the development of motive power from such a source (1673) but it was not until 1860 that Etienne Lenoir produced a workable gas engine with electric ignition. In the meantime, about 1820, something like a "modern" internal combustion engine had been schemed and even made in a laboratory form by W. Cecil, one of those remarkable English parsons of the period who applied themselves to physics while (usually) their curates read the Offices in Church. He used hydrogen mixed with air, which must have needed considerable faith in divine protection. In 1838, W. Barnett struck upon the idea of compressing the mixture before firing it, and their work was of first importance to Lenoir in the production of a commercial, and still stationary, gas engine. Improvements continued through Nikolaus Otto and Eugen Langen. Otto produced the four-stroke engine about 1876, and Sir Dugald Clerk, a Scotsman, made the first two-stroke engine in 1881. Petroleum spirit was the fuel used by the great motor-car pioneers Gottlieb Daimler (1834-1900) and Karl Benz (1844-1929). During the 'eighties, first Herbert Ackroyd-Stuart in England, and then Rudolf Diesel in Germany, devoted

The Westinghouse locomotive that first hauled mainline trains, Baltimore, 1895.

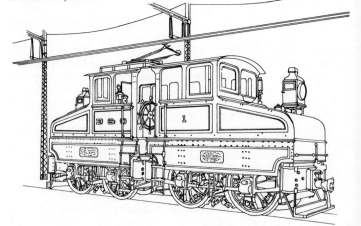

their energies to the production of stationary engines which would consume crude oil with ignition by compression. Both produced satisfactory stationary engines, by the standards of the time, but it was long before compression-ignition oil engines were to furnish locomotion. One notes at once that Benz gave his name to refined petroleum for fuel and Diesel to the compression-ignition engine we now know so well.

The history of internal-combustion engines is a study in itself, like that of steam in the eighteenth century. Suffice it now, that in 1892, a very little industrial railway locomotive, powered by a Daimler engine, was built by the Esslingen Engine Works, Emil Kessler's old foundation which had built steam under Crampton's patents some forty years before. She is on p. 186—necessarily somewhat sectioned—with her double-buffer arrangement for moving both railway wagons and factory trams! Few could have guessed, either then or at the end of the century some eight years on, that by the nineteen-sixties internal combustion of one sort or another would be moving almost the entire overland transport of North America, and of many other parts of our long-suffering planet! As yet, straight steam had a long way to go!

Self-contained steam car on the Pilatusbahn, Switzerland, 1889.

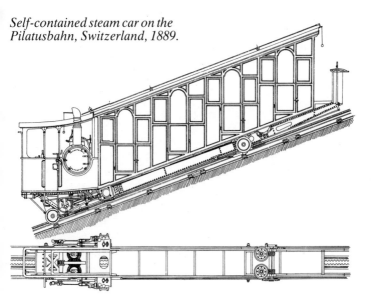

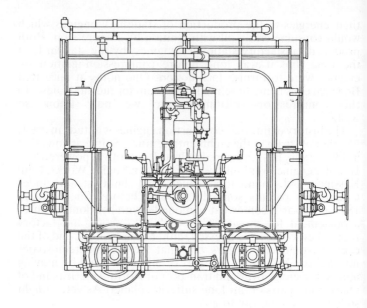

The little Daimler motor-
locomotive built at the
Esslingen Works, 1892.

Chapter 6
Steam Challenged

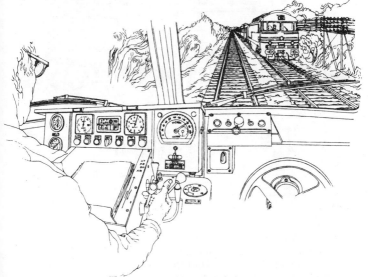

In England, birthplace of the commercial steam railroad, people fixed its conventional beginning to the year 1825, whatever may have happened in 1804 or 1812. The opening of the Stockton and Darlington Railway was taken to be "Milepost Zero". In 1875, there had been quite a ceremony over the *Railway Jubilee*. Nobody bothered much about the seventy-fifth anniversary at the turn of the century. In 1925, there was a tremendous celebration of a Railway Centenary up at Darlington, with the Duke and Duchess of York (later, *malgré eux*, to be King and Queen) as Guests of Honour, the London and North Eastern Railway Company running the show, and enthusiastic support from Belgium, Italy and Ireland who sent ancient locomotives for exhibition, or, in the case of the Italians, presented a magnificent iron plaque honouring George Stephenson.

In the last quarter of this Railway Century, the steam locomotive had really begun to encounter its mechanical challengers. On the rail, it was already fighting a rearguard action against electric traction, not only in city railways but in mountain countries where a great fall of water produced

187

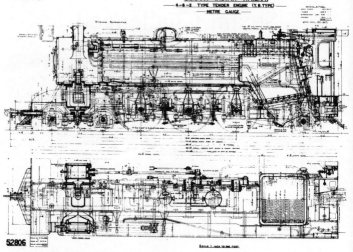

*The art of locomotive draughts-
manship: the drawing for a South
Indian Railway 4-6-2 type tender
engine with piston valves. Made by
Nasmyth Wilson & Company, Manchester, 1928.*

corresponding something-for-nothing (subject, of course, to
immense capital costs and reasonable maintenance thereafter).
In those years, too, the compression-ignition oil engine
emerged, a difficult child at first, after equally difficult travail.
Further, and not on the rail, but with a here-I-am flourish on the
old roads, there had arrived the mass-produced motor-car,
beginning with Henry Ford's first T-Model. But in those years
1900-25, the steam railway-engine was yet a lusty old buffer,
and was to stay so, in many places, for yet another half-century.

There were improvements and new inventions galore. There
was the limited burning of oil instead of coal for steam raising,
pioneered in Russia by that ingenious Scot Thomas Urquhart,
and furthered, though to a limited degree, in Pacific-Coast
America and East-Anglian England. The thing that really
mattered during this time was the super-heating of steam in an
otherwise entirely orthodox locomotive, and for that we must
thank, for all the years over which it has benefited us, Wilhelm
Schmidt. Allow some credit, also, to the old Royal Prussian

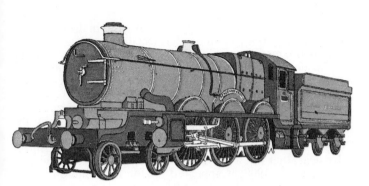

Great Western Railway's Castle *class 4-6-0, built in 1923. Shown is* Pendennis Castle.

State Railways, the K.P.E.V., which believed in it, and built the apparatus, and showed it to be good!

The idea of drying live steam on its way from the boiler to the cylinders was ancient. The point had been the question of how to increase its temperature, and therefore its efficiency in those cylinders. Old English ideas embraced a drying chamber in the smokebox, but that was not enough. In France, Jules Petiet had put it on top. Where was the real heat? In the firebox indeed, but that had business enough! Where next? In the flues! Therefore, by Schmidt's physics, let the steam be taken through the flues! How, in the name of all that was impossible?

The answer, said Schmidt after some thought, was quite simple!

At least, so it seems in retrospect, though there was long experiment with all the toils of trial and error. The basic arrangement of the Schmidt superheater, and of the numerous other apparatus in the same family, which came later, was this. Instead of the live steam passing straight from the dome or other boiler steam space to the steamchest and thence through the valves into the cylinders, the steampipe was divided, its upper portion leading to a header whence a large number of small steam pipes or *elements* were led into, and again out from, a corresponding set of enlarged flues in the upper part of the boiler. Each element followed the line of a woman's hairpin. The live steam, passing through these, was thus

189

subjected to the fierce heat of the firebox gases in the large flues before passing to a second header, usually combined with the first, whence main steampipes took it to the steamchests. At the first header, the steam was *saturated*. At the second it was *superheated*. The stuff was, in Richard E. Trevithick's happy phrase, *strong steam*.

In 1898, the first two locomotives to be in regular service with the Schmidt superheater appeared, as suggested, on the Prussian State system. The apparatus was applied to Prussian Class P 4^1, a 4-4-0 type with outside cylinders and Heusinger valve gear, that was a characteristic German express engine of the period. In 1900, it appeared for the first time on a tank locomotive, again a Prussian State 4-4-0 (Class T 5^2), for the Berlin Metropolitan Railway. In this, an enlarged and much extended smokebox was used, and this came to be a usual complement to the Schmidt and kindred superheaters, which spread through Continental Europe, the British Isles and the Americas from about the middle of the succeeding decade.

For the next quarter-, indeed the next half-century, the orthodox steam reciprocating railway locomotive was to be distinguished apart from increasing size by the use of the superheater, steadily increasing pressure, generous firebox space, and the supersession of link motions by radial valve gears of the Walschaerts/Heusinger sort. As working pressures rose, and especially with the introduction of superheating, piston valves came to replace slide valves for steam distribution though even those were of some antiquity. They had appeared prematurely and disastrously in the England of the 'seventies. France in 1884 was more fortunate; on her old State Railway, Ricour produced successful piston valves with air-admission valves. By 1910, piston valves were *in*. For a locomotive of this phase we exemplify in a coloured chart a characteristic British-built locomotive for the South Indian Railway, additionally interesting in that it shows a relatively large locomotive on one of the narrower gauges (one metre).

Gauge variations we have mentioned before. Though the greatest of all rail-gauges, the generous 7 ft. of the Great Western Railway in England, had come to an end in 1892, variety continued. North America, Mexico, and Western Europe apart from Ireland, Spain and Portugal, had long settled on the classic Stephenson gauge of 4 ft. 8½ in., or 1.435 m. Doubtless out of strategic considerations, Imperial Russia adopted a gauge of 1.524 m. (five English feet) which persists in the Soviet Union from its western borders (including Finland) to the Far East. China, however, has the

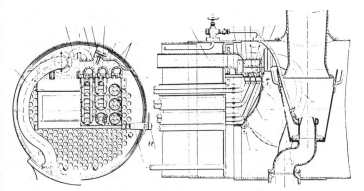

*Schmidt's superheater, 1898, became
extensively used on steam locomotives
throughout the world.*

Europe-America gauge. The broadest gauge in general use has
long been that of 1.676 m. (5 ft. 6 in.), which is that of India,
Pakistan and Ceylon, Spain, Portugal, Argentina and Chile.
All these lands apart from the Iberian Peninsula were long
under British railroading influence. The 5 ft. 3 in. gauge (1.600
m.) is that of Ireland, Brazil and the southernmost parts of
Continental Australia. The last-mentioned country, however,
had (in other people's eyes and latterly in its own) about as
cock-eyed a gauge policy as any country in the world;
early-Victorian England, pre-Civil-War America, and Sweden
included.

In the last century, the several Australian States were full of
jealous small-nationalism. Customs barriers were set up, and
the several States had their own rail-gauges, usually chosen to
be different from anything next-door. Thus Victoria and South
Australia chose the Irish gauge for their main lines, New South
Wales chose the European and American standard, while
Queensland in the East, and Western Australia at the other end
of the Continent, chose the 3 ft. 6 in. gauge (1.067 m.) which
Tasmania also used, not to mention New Zealand and the great
systems of South Africa, Japan and the then Dutch East Indian
islands. But Australia has long rued her legacy of mixed gauges.
In 1915, even, the Trans-Australian Railway was completed
with largely New South Wales equipment, making two more
breaks. Further, South Australia used the 3 ft. 6 in. gauge for
certain up-country lines, as well as the broad gauge on main and
inter-city lines. As we write, there are still three gauges

191

in Port Pirie station, South Australia, a situation we remember also in Växjö, Sweden, and on the Franco-Spanish transfer at Hendaye, whence, Gustavo Reder remarks, begins the largest metre-gauge network in Europe, even bigger than the Swiss, and serving Pullman-type express services west of San Sebastian.

The 3 ft. 6 in. gauge (1.067 m.) was the third one at Växjö, as was the metre gauge at the end of France. It is a surprisingly adequate gauge for mountainous countries (Norway once ran expresses to Trondhjem on it) and today South African trains are otherwise the dimensional equals, or even superiors, of many in Europe, with great locomotives that have even impressed visiting Americans. Metre gauge has been much used for secondary lines in India, and in Malaysia and East Africa also. Many narrower gauges have been used about the world since the first public narrow-gauge railway was opened between Portmadoc and Blaenau-ffestiniog in North Wales, away back in 1836. Its gauge was one of two English feet, which latterly became the rather curious one of 1 ft. 11 in. The widely-used Swedish narrow gauge of 0.891 m. has stirred curiosity in other countries. Why such a gauge? Or so ask both Americans, and other Europeans, for by English measure it is 2 ft. $11\frac{25}{32}$ in.! Quite simple, though! It comes to exactly three Swedish feet and that is a solution pleasing to any artist previously puzzled by figures.

In this first quarter of the twentieth century, the world's railway map became almost complete. Great estuaries had been bridged as man had not previously imagined, and train-ferry ships had spanned similar, and much greater gaps. Both these things began in Scotland on the Forth and Tay estuaries where still one may see some of the greatest bridges in the world. North America was long spanned, and several times over, by thin, steel-blue tracks. Russia and Japan both went for Manchuria, whose crossing Russia won at the cost of a subsequent, rather humiliating war. On January 1, 1903, by dint of the Trans-Siberian and the Chinese Eastern Railway (which also was Russian-owned, on the Russian gauge which was so minutely over-standard) it became possible to travel by train, or rather trains, from Calais on the Straits of Dover (*La Manche* to certain old and well-loved friends) to Vladivostok on the Sea of Japan in rather more than a fortnight. There was a train-ferry crossing on Lake Baikal, elimination of which was hastened by the immediately impending Russo-Japanese War. At the other end, the Channel Tunnel had been talked of since the eighteen-seventies, but as yet there were not even train-

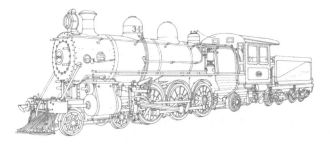

New Zealand Railways Class Q, built in 1901.

ferries, though an Englishman, that unhappy genius Sir Thomas Bouch who built the first, disastrous Tay Bridge in Scotland, had been their father. Later, too, came the very circuitous all-Russian route to Vladivostok by the Ussuri line through Khabarovsk.

Nearly complete was the railroad map of the world, but it was still almost entirely a steam realm, and that brings us back to the development of the ever-bigger steam locomotive engine.

Firstly; to the fast passenger locomotive! We have seen the arrival of the Atlantic type from America. We have seen that the 4-6-2 type had appeared in the United States quite accidentally, owing to a matter of axle-load restriction with the heavy ten-wheeler, plus the needs of experiments, in another case, under the Strong Patents. The true Pacific, the elongated Atlantic with three coupled axles, was also American in origin but not in application.

2-8-2 Mikado-type locomotive for Nippon Railways, in 1897.

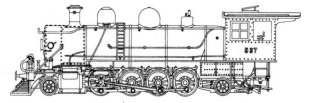

Baldwins built it in the States; but first for New Zealand, which Americans then regarded as a British colony, whatever New Zealanders might think. (It was then generally English down to the middle of South Island, and extremely Scottish at Dunedin and all points south.) The New Zealand Pacific type engines were in every way American; further, they anticipated on a small scale what was to be the conventional fast passenger locomotive in the States and in Canada for the next twenty-five years, with a big firebox overlapping the main frames and supported by a small pair of trailing wheels. The real object was to allow for a wide grate burning low-grade New Zealand coal.

Akin to the 4-6-2, but originally intended for fast freight haulage, was the equally American 2-8-2 engine, and here again it was first built for export before running in its native continent. Japan had the first examples, hence the American type-name Mikado. Over forty years later, the type was still being built in immense numbers for American military use, and then for post-war replacement service.

But at this time, at the beginning of the new century, there were other attempts to achieve a steam locomotive steady in running yet with plenty of space for a big firebox. None was less conventional, yet at the same time workable, than that of Giuseppe Zara, using Plancher's system of compound expansion, on the Italian Southern (Adriatic System) Railways, from 1900 onwards.

Our drawing is fairly explanatory. An old-fashioned Irishman would say that the front-end was at the back and, accepting that term, he would be entirely right. Having the boiler reversed allowed for ample firebox space over the bogie, and also placed the cab in front, with a free view of the road ahead. It was a most beneficial arrangement for long tunnel

*The "cab-forward" 670 class 4-6-0 was
designed for the Southern Adriatic
Railway in Italy by Giuseppe Zara in
1900.*

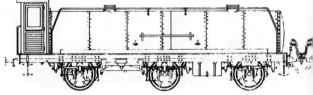

working. The engine was four-cylinder compound. Its only serious disadvantage was the need for side coal bunkers, with consequently uncomfortable firing. The long cylindrical tender was for water only. The type was sometimes called *Mucca* (not quite a compliment), and one of the first (No. 3701) was shown at the Paris Universal Exhibition of 1900, a wonderful show of contemporary locomotives, of which this was surely the most advanced by the standards of the time.

Many more of the "Mucca" type were built in the next few years. We saw them working some of the best expresses between Milan and Bologna in 1922, and over ten years later they were still serving those on the Milan-Verona-Venice run, a remarkable achievement for what was regarded as a rather eccentric design.

Later, in the United States, the Southern Pacific Railroad built enormous "back-to-front" locomotives, on the Mallet semi-articulated plan, for heavy pulling on the slopes of the Sierra, and in these the use of oil as fuel ruled out the old disadvantage of side bunkers.

Placing the cab in front was, of course, one of the great advantages of the electric locomotive which was now on its cautious way in, as of the oil-electric locomotive in later years, but there were several more attempts to do it with orthodox steam locomotives. In France there was that of Thuile, whose engine was also exhibited in Paris in 1900, of which the less remembered the better. She turned over on her trials and killed her unfortunate designer, so that was that. She was remarkable in having 7 ft. 4½ in. coupled wheels. Henschel and Son of Kassel, makers of so many memorable German locomotives down the years, built two remarkable cab-in-front locomotives in 1904. One, a 4-4-4 (2-B-2) express engine, the Prussian State classified S 9, had coachlike casings with side windows

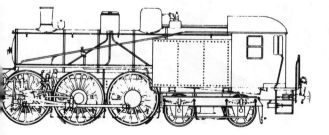

over both engine and tender, with a pointed front, making her look like an engine with an electric "top" (apart from the protruding chimney) and a steam "bottom", cylinders, valve-gear and all. Firing her was an abominably hot business. The other was a 4-6-4 tank engine with a cab at each end (Prussian Class T 16) with a double set of controls but, of course, with the fireman at the firebox-and-bunker end. The Prussian 4-4-4 by Wittfeld is sometimes instanced as the oldest example of a streamlined steam locomotive, though in France Baudry, with his *coupe-vent* engines, was aiming at that, and the idea was older still, going back at least to a completely streamlined experimental train on the French State Railways, by Ricour in 1887, using Etat locomotive No. 2072.

Still, the steam locomotive was undisputed Empress of the World's Railroads, as some American *siderodromophile*—or perhaps business-booster—might have put it. (That word, which we believe we invented, is quite good Greek, though not Classical Attic!) The less ornate American word of our day is *railfan*. Germans in particular regarded the electric locomotive, which their fatherland had sired, as a subject for laboratory treatment. Trials were held on a military railway, Marienfelde-Zossen, with an experimental locomotive and two equally experimental motor cars in 1901 and 1903. Three-phase traction was used. In 1901, the top speed attained, with the Siemens and Halske locomotive, was 101 miles an hour. In 1903, both cars attained 130 miles an hour, which remained a rail-speed record until 1931. Observers

Prussian State Railways (Class S9), 1904.

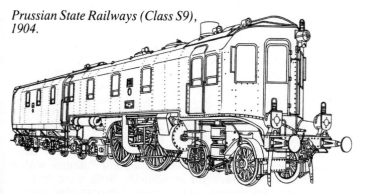

(Opposite) *Detail of "Clapham Junction" by Terence Cuneo.*

noted the noisy impact of pebbles against the car floor. We have known people incredulous about this, but we had an isolated experience of it, lasting a few seconds only, with steam traction on the London and North Eastern Railway in 1935. The speed at that time was just over 112 miles an hour. The phenomenon depends on air currents and the state of the roadbed.

On the Great Western Railway in England, with a special mail train from Plymouth to London, Charles Rous-Marten recorded a maximum speed of 102 miles an hour (plus a fraction) while running downhill east of Exeter, on May 9, 1904. More importantly, on that run the initial engine, *City of Truro*, ran the fifty-two miles from Plymouth (North Road, pass) to Exeter (St. Davids, pass), over a quasi-mountainous route, in a fraction under fifty-six minutes, and averaged just over seventy miles an hour, pass to stop, from passing Exeter to stopping outside Bristol. Thence to London (Paddington) with the train lightened to a mere 120 tons of mail cars, the engine *Duke of Connaught* took just over ninety-nine minutes for the stretch of 118.5 miles. This was a run in regular public service, not a trial for show, as in the case of New York Central No. 999 a decade before. Learned attempts have been made to disprove both claims, but without convincing success. It should be added that at the time, in England, the Great Western company was in fierce competition with the London and South Western for the Plymouth traffic, and very energetic running was made by both companies then and for several years after. They culminated with an accident, as had the old races between London and the Scottish cities. The South Western company spilt its train all over the Salisbury curve in 1906, with heavy loss of life, and that was followed by a *working agreement* between the interested companies. Actual racing by rival railway companies between the same cities was a prank peculiar, to our knowledge, to the British and the Americans. What is remarkable is that the locomotives employed by the Great Western were of relatively archaic type. Both had inside cylinders and double frames, stemming from the Stephensons and their contemporaries long ago. Neither was particularly large, though *City of Truro* had a generous boiler with a Belpaire firebox. *Duke of Connaught* had very large single driving wheels; she was *almost* a giant Patentee with a leading four-wheel truck for steadiness. Both were of fine, ornate, Victorianly-brassy aspect, quite different from rugged Pacifics and suchlike!

Still as industry and liberal capitalism expanded, the mass

movement of freight, fuel, ore and passengers spread with it. Vehicles grew heavier and trains larger. In the States, the mighty Pennsylvania Railroad released a photograph, doubtless with pride, of three 2-8-0 locomotives toiling hard with sixty coal cars. In passenger service, ordinary cars grew more elaborate, apart from the expensively luxurious vehicles of such companies as Pullman and Wagons-Lits. Travel in an American chair-car from Chicago to St. Louis, or in a good English third-class carriage from London to Plymouth via Bristol, fairly comparable in distance and time, could be not just tolerable but agreeable, on plush and quite elegant surroundings, and while both Englishmen and Americans were supercilious about the hard third-class of Continental Europe Germany and Scandinavia in particular furnished a very superior second-class at quite cheap rates. Second-class was extinct in Scotland and moribund in England. In Democrat-Republican America, "class" was as dirty a word, still, as it was to be in Russia after 1917. Realist Germany under the last of the Emperors made tremendous use of fourth-class which, in our own experience, was all right for journeys of necessity or for spending a jolly Sunday among mountains. But all these things made for heavy trains, with bigger and yet bigger engines.

Queen of steam locomotives for fast passenger trains, in North America and Western Europe during this quarter century, was the Pacific, the real Pacific, 4-6-2 with long frames and ample firebox, whose prototype we saw built by America for New Zealand at the beginning of the century. Her kind was a classic. In the early nineteen-hundreds, canny American companies, which nevertheless worked their equipment to death and then bought new, took it up. Let us take our first instance from the Pennsylvania Railroad, though one seems to remember that the Chicago and Alton showed the way.

On June 11, 1905, the Pennsylvania company revived its famous Pennsylvania Special train, competing with the Vanderbilts' New York Central Lines, and scheduled it to link New York and Chicago, both ways, in eighteen hours. That meant an average speed, including stops, of 50.2 miles an hour, which by the standards of the day was extremely creditable. It entailed an average speed of 57.8 miles an hour over the first 189 miles from the old Jersey City Terminal to Harrisburg, with one intermediate stop, which was still more creditable. The engines at that stage headed the train as a couple of stately Atlantics, or even as one over the initial level section, e.g. Class E3d with piston valves.

THE PACIFIC

In 1905, the Pennsylvania company invested in its first Pacific, an experimental engine classified K 1, from the American Locomotive Company (the "Alco" of later years). In 1910, Class K 2 was designed and built, in increasing numbers, from 1911 onwards. This had the big, broad Belpaire firebox. Modified for superheater, in 1913, the design developed into Class K 3. By that time, the Pacific had generally arrived in North America.

In Europe it arrived during the two years 1907-8. France built it, initially in the very successful and handsome Paris-Orleans No. 4501, a de Glehn-du Bousquet compound by the Societé Alsacienne of Belfort. Then, under that great man Anton Hammel, the German firm of J. A. Maffei built it for the Badenese State Railways. In 1908, another giant of locomotive design, George Jackson Churchward, built a solitary example for the Great Western Railway, whereon it was something of a white elephant, being barred west of Bristol owing to insufficient clearances. (The boiler, too, was too long between tube-plates.) Churchward, indeed, had already produced a highly efficient type of four-cylinder simple 4-6-0 engines which, with successive enlargements, were to suffice his company to the end of its days in 1957. (Pacifics were not to become familiar in Great Britain until the nineteen-twenties.)

To Hammel of Maffei's works in Munich belongs the credit for producing the perhaps finest of these early European Pacific-type locomotives, and certainly one of the most beautiful. This, originally Bavarian State Railways class S 3/6, was an enlarged, superheated, big-fireboxed version of Hammel's S 3/5 Atlantic and, even more nearly, of an

City of Truro,
1903.

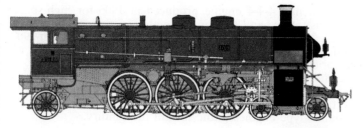

(Above) *J. A. Maffei built this express 4-6-2 steam locomotive Type S 3/6 in 1923 for the Bavarian State Railways.*

(Centre) *The 2-8-2 Mikado type passenger locomotive, Type 5, was built for the Belgian National Railways.*

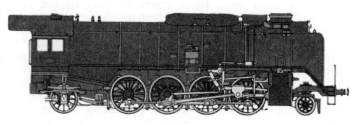

experimental 4-4-4 engine produced in 1906 for high-speed demonstration, and now in Nuremberg Transport Museum. The S 3/6 was four-cylinder compound, produced first in 1908 and last in 1931, with but little variation down the years, an historical circumstance with little parallel in the history of steam-locomotive design. That design in itself was a remarkable marriage of European and American tradition, certainly influenced by Bavarian purchase of a pair of Baldwin Atlantic-type engines at the turn of the century.

It was not a big engine by American or later European standards, as Pacifics went, but down many years it could handle with no apparent effort anything from a heavy international express to a Sunday excursion train between Munich and the mountains. Visually it was most beautiful, a study in visible as well as of mechanical balance. The last examples were in service until 1965. Two have been preserved; one in the German Museum, Munich, and another in Switzerland. To many Germans they were *die schönste Lokomotiven der Welt*.

Yet for all this impact of so splendid a locomotive type, Europe was shy. France, to be sure, received it with joy; all her great regional companies took it up. North Germany would have none of it until the middle nineteen-twenties. Sweden alone in Scandinavia produced it (Swedish State Class F) in 1914. Like the Bavarians, these were four-cylinder compound engines with the low-pressure cylinders outside (and in this case sharply inclined). The Swedish engines, later displaced through electrification, went to Denmark, as a result of which the class was to last over half a century. The French Pacifics, both simple and compound, did some of the finest fast passenger work in the world over many years. Largest and most powerful of the older European Pacifics were those of J. B. Flamme on the Belgian State Railways. Some of these also survived for upwards of fifty years, though others had been lost by war when still quite new.

Italy cautiously took up the Pacific, and Russia and Hungary even more cautiously. These countries were to use very extensively the 2-6-2 type of express engine, which the Middle-Western United States were already calling the Prairie type, and at moderately high speeds it served them well over many years of steam traction. Austria and Hungary both made use of it. Hungary built 950 from 1909 onwards, plus twenty for Slovakia in 1942, all of one class (324) and 105 of other classes. One should record here that it was the success of the Italian and Austrian engines that led to the production of the very successful, and in our opinion very handsome, Russian Class Su, which continued to be built long after the Revolution.

Russia at that time, before 1914, made much use of the Mogul engine, the 2-6-0, and so did many other European states, though it was beginning to be regarded as effete and old-fashioned in North America where freight trains in particular were growing heavy beyond the worst nightmares of old-time locomotive men.

Prussia, Bavaria, Italy, all the Scandinavian countries, and to an increasing extent England were finding it an extremely useful locomotive for both the lighter freight and the slower passenger trains, and Italy in particular had produced in her State Railways' once-famous 640 class what was for a while a highly satisfactory express passenger locomotive, to be seen even in the nineteen-sixties on the level lines in the North, still under steam traction in a much-electrified land. Italy's first superheater type, it was an admirable locomotive for moderate traffic, whereon it lasted on more lightly laid lines in both the Americas (usually either English- or Scotch-built in the South American republics, most of whose railroad systems were British-owned throughout the Steam Age). For a curiosity: England's only two Moguls, over more than a decade, were two of a very English design which South America, Australia, Spain, and several other parts of the world knew well enough. They went to a rather remote English company, the Midland and South Western Junction, which had lately fallen on

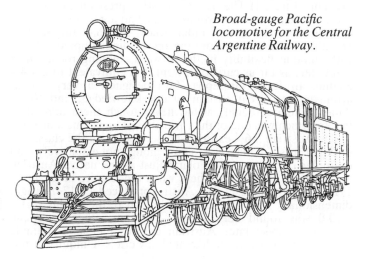

Broad-gauge Pacific locomotive for the Central Argentine Railway.

evil days but had recovered under brilliant management, and the first of them came to it owing to some other insolvency in South America. Countries with railways under company management sometimes saw this sort of thing; State railways rarely went bargain-hunting, though now and then they sold their obsolete equipment to smaller and less opulent outfits. Even the British Government, long years after, managed to *flog* some ageing, surplus War Department locomotives to the Shanghai-Hangchow-Ningpo Railway, and the same sort of thing was to happen, from both the British and the American Governments, when the fighting stopped in 1945, still longer years after!

In Austria, at this time as at others, there was unorthodoxy, not of an idle sort. That great man Karl Gölsdorf was no longer as young as once but he was far from being *vieux jeu* in his work and perhaps only he would have produced what we call, in all humility, the *reversed Pacific*. It perpetuated the Italian Zara's "Mucca" idea of putting the great firebox over the bogie. But his engine was nevertheless front-end-forward, for she was 2-6-4, still a four-cylinder compound.

She had the big cone-barrelled boiler which Churchward was using with great success on the Great Western Railway in England, and at which other British locomotive men were looking with much distrust, for it put them in mind of the modified American wagon-top type. (Gölsdorf of Austria and Churchward of England certainly visited each other, and exchanged notes!) The new Gölsdorf express engine of 1910 was 2-6-4, but under her front-end she had not that mechanical abomination, the Bissell radial truck. She had the Krauss-Helmholtz bogie, incorporating the leading coupled wheels, with suitable flexibility in the forward coupling rods, which made her, as far as curvature went, the equivalent of a 4-4-4 engine. In Italy, Giuseppe Zara designed a very similar arrangement. Great Britain to our knowledge, and North America to our belief, would never touch these admirable trucks, which was one of the curiosities of mechanical history.

The Austrian 2-6-4 express engine was one of the classics of its day. Prussia made a close copy, though not extensively, and examples of this went to Poland, just as the native Austrian article found its way elsewhere when the Empire broke up in 1918. For mountain service, Austria produced a 2-12-0 design, dimensionally akin to the 2-6-4, but on traffic of this sort a 2-10-0 was found adequate, as in many other European countries. These later Austrian engines had an inimitable elegance of their own, unlike anybody else's. Uniquity of this kind could be claimed for several styles of this period—those of the Great Western in England, of the Pennsylvania in North America, of Maffei in Munich and of the Northern Railway of France—but not for many.

Before we turn to freight haulage, we should consider passenger conveyance. In one of its least exciting yet most important forms, this involved the transport of immense numbers of people daily in and out of the world's largest cities. Some of these, such as London, Berlin, Hamburg, Paris, and rather surprisingly Chicago, saw great use of the side-door coach, either with compartments or, as in the American cities, variations on the compartment plan with long cross benches. Chicago, be it added, had found their advantage in traffic to and from the World Fair of 1893, and for some years, the Illinois Central Company standardized side-door cars for what America had come to call *commuter traffic*.

As that term has come into general use in the English-speaking countries to denote twice-daily movement of office workers in city areas, whatever the form of transport, let it be at once explained. A commuter was simply a person who, by virtue of making regular journeys by public transport, had his fare *commuted* to a much reduced sum payable by periodic contract; weekly, monthly, quarterly or even annually. In the phraseology of Southern England he was a *season-ticket holder*; in that of Northern England he was a *contract-holder*; in America he was a *commuter*. The term "commuting motorist" for a man who drives himself to and from town,

The Tait suburban second-class car, 1911, for use in Melbourne, Australia. It seated 92.

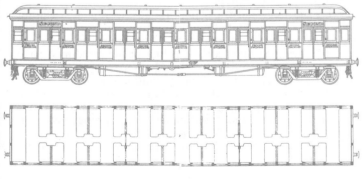

which is regrettably used in many areas now, is a terminological absurdity.

The American commuters' car was generally the same as an ordinary American day-coach, extremely difficult of egress and access at terminals. In the early nineteen-hundreds, certain American designers added middle double sliding doors to help things out. Boston cars, thus equipped, were closely copied in England by the District Railway on its electrification in 1905. The District company was just a little too clever. With the best intentions, it tried to make the doors automatic, with opening and closing by little air engines worked off the main brake-air supply. The things speedily caught in an inexorable grip people's arms and legs, as well as old gentlemen's beards, ladies' long skirts, and other appendages, with variously shocking results. This first attempt at automatic doors, now so familiar, lasted some six weeks, and then it was subjected, to your author's knowledge, to an official Act of Oblivion. For many years after, the thing was unmentionable in London Transport circles. A remotely-remembered Peter Arno cartoon ("By God, suh! I won't forget this insult!") suggests that America also had troubles of this sort, even more thoroughly wiped off the slate of history. Early London underground cars, through the influence of Charles Tyson Yerkes, were extremely American, and very good cars at that, except when the sliding doors were left open on a winter night. On hot summer nights, passengers inclined to leave them open a-purpose, except when the cars were so crammed that the passengers would inevitably spill themselves out on the tracks and get irreparably damaged.

The Illinois Central cars followed American contours, and had sliding doors, as did, a little later, the Australian commuters' cars far south in Melbourne. We show here a Tait suburban car—prototype of these—from a diagram book of 1911. The very narrow gangway was to enable rush-hour passengers to distribute themselves up-and-down the car after their first scrimmage at the doors. Cars of this type, and also of the old swinging-door sort with that peculiarly Anglo-Saxon abomination, the full-width closed compartment, were later converted for electric working in multiple-unit.

In fairness to your author's own race, let it be added that those closed compartments were common enough at the time in North Germany, France and many other European countries from the Peninsula to the Peloponnesus. They had an awful habit of turning up on the less respectable trains over longer distances, especially on cheap excursions, where their frequent lack of water-closets originated all sorts of curious stories, amusing to those who had not themselves suffered. Alas, not so long before, this defect had distinguished nearly all Western European trains! Blessings on those nameless Americans and Russians who first made suitably-adorned holes in car floors! Germany showed up well, from the 'eighties onwards.

PASSENGER CLASSES

Throughout the world, there remained two conflicting ideals of the railroad passenger car; the "open" type chiefly favoured by North America, Russia, South-west Germany, much of Eastern Europe and Scandinavia, and the compartment-form of Western Europe and of most countries under British influence including India but, rather surprisingly excluding New Zealand, which, where railroads were concerned, was decidedly American in practice. Both types had their earnest protagonists. Abroad from America, Mark Twain lamented his isolation, and possible immolation with a troublesome drunk. Abroad from England, Rudyard Kipling wrote of an Egyptian train: that she reminded him of a South African train and that consequently he loved her, adding that the familiar arrangement of compartments served by a side corridor equally annoyed his American companions (who may have included Mrs. Kipling).

So far, passenger cars had been very generally made of timber, at least as to the bogies, which made them comfortable by good insulation but liable to burn-up in a wreck when lit by gas or kerosene. One recalls awful classics of this brand of smash from Ashtabula in the States, Gretna in Scotland, Bellinzona in Switzerland and a good many other places before and within our time. To be fair to the Swiss, whose fault it was

Reserved passenger car built for the Nitrate Railways of Chile, 1927.

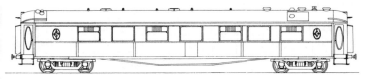

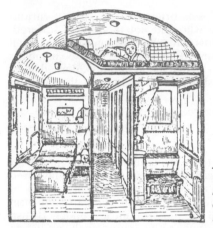

A German anticipation (1924) of the later Pullman roomette. Its second-class passengers slept in the roof.

that their trains collided, it was a German gaslit coach that burst into flame and instantly affected an Italian wooden one.

In the United States, land of cut-throat competition, the more wealthy companies answered public scare over these things first by building much more substantial cars, almost invariably to what one may conveniently call the Pullman outline, and then making them entirely of steel. Less wealthy companies sometimes covered their existing wooden cars with thin steel plates, studded by ostentatiously massive rivets, to kid their patrons into believing that this was indeed an all-steel car. "The all-steel Columbian" was a publicity-department definition of the period, by the Chicago, Milwaukee and St. Paul company, whose other crack train was called the Olympian.

Some of these old American transcontinental trains were of quite astounding sumptuosity. A barber-shop was a sort of status-symbol, and so was the observation car at the rear,

Plan of a private car called "The Nomad", typical of the luxury on American railroads in the early days. It ran on the Rio Grande and Western Railroad.

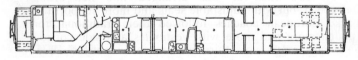

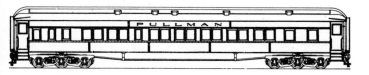

The great steel Pullman car character-istic of North America in 1908-28.

whereon one could either sit out in a folding chair on the back platform, watch the scenery, and put up with the dust, or sit on a handsome plush one inside with the *Wall Street Journal* and let the scenery take care of itself. The Milwaukee company, boosting its electrified Mountain Division, reminded its patrons that they could sit outside *with no smoke to annoy, or obscure the view* (cf. *The King of the Rails*, 1915). They were lovely things, those old American railroad brochures! North of the 49th Parallel, in the Canadian Rockies, the Canadian Pacific Railway was already using a form of glazed observation car which anticipated the "dome cars" of both Canada and the States in later years. Remote origins of the arrangement were in Russia, under Winans' influence, back in the eighteen-sixties.

American passenger cars at this time were almost invariably of the clerestoried shape, still favoured also by Prussia and Scandinavia, the Midland Railway in England, and (where it was to last longest) in South Africa.

Passenger accommodation was very variable indeed, according to the country and the sort of user. The most comfortable ordinary arrangement for daytime travel through-out North America was the *parlor car*, a Pullman or Pullman-type vehicle with superior overstuffed chairs on fixed pivots, set singly each side of the central aisle. It had been used on several British railways, chiefly in the South, since Pullmans were first imported in the middle-seventies, but in 1908 came the first British-built Pullmans, still with single chair seating, but with the very massive and comfortable chairs movable on legs.

America's chair-car was arranged like her ordinary day-coach, but with adjustable seats and more support to the back. European coaches followed their traditional form, though more and more with vestibules and side corridors. England's remarkable third-class we have noticed already.

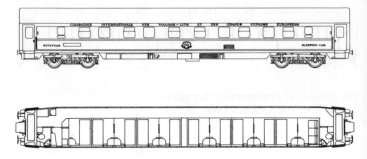

A modern wagons-lits sleeping car.

The farther east one went, from Europe into Asia, or south into Africa, the cheaper became the third-class fares and the more austere the third-class carriages; also in some countries the dirtier they were. A colleague told us, of Europe in the 'thirties: "The bugs begin east of Budapest!" Highest standards of European cleanliness were undoubtedly in Scandinavia, where, also, the Swedish State Railways introduced third-class sleepers in 1910, with three superimposed berths to each compartment. Some years later, these were copied in Germany, and later still in France. For all the excellence of her third-class day coaches, England held off third-class sleepers until 1928, when they took the form of four-berth compartments, sleeping two each side as in a rather indifferent second-best-first-class arrangement which the French called *compartiments à couchettes*.

On an ascending price scale, American Pullmans contained many sorts of sleeping compartment, with side-corridor access, in addition to the traditional open section dating back to 1865. A *bedroom* resembled a European two-berth Wagons-Lits. A *drawing room* was more spacious. A *master-room* was fairly plutocratic.

With the general introduction of steel cars, design became highly standardized. A Pullman berth on the Florida East Coast, rolling out over the old viaducts to Key West, half-way to Cuba, was exactly the same as a Pullman berth on the Great Northern, headed for British Columbia, green plush and all. D. H. Lawrence found the same thing in Mexico, though with a whiff of bloody-handed brigands somewhere beyond its close-shaded windows. In Europe, you could not possibly mistake a car of the Holland Railway for one of the Prussian

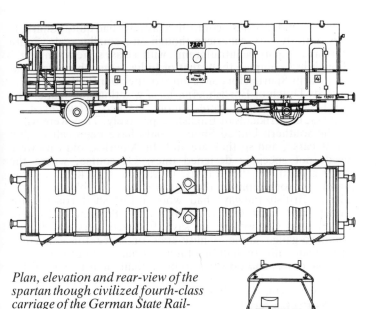

Plan, elevation and rear-view of the spartan though civilized fourth-class carriage of the German State Railways.

State, or either for one of the London and North Western, though most of them were of the primeval side-door type in the earlier part of this century. Some of the Danish State carriages were more like those of the Italian Mediterranean Railway than those of the Hässlehom-Hälsingborg just across the water, but still very different. Even the mighty and far-reaching Wagons-Lits company, the nearest approach to America's Pullman undertaking, was lusciously variable.

Eastern third-class carriages were fairly fearsome, quite apart from the fact that they were invariably crammed. In India, each long compartment had four wooden benches, the middle ones back-to-back, with access to a small Oriental-type latrine. The late Mr. Gandhi, with all India at his feet, always went third-class as a man of the people. Rich Indians, and official and Army English, travelled first-class in dusty luxury. Babu-class Indians went second-class. Some Eurasians and *poor-whites* used a thing called intermediate-class, lower than second, and in Kipling's words, "very awful indeed".

Japan, and Japanese-controlled railways had quite admirable cars, latterly very American though on a smaller scale. South Africa had many beautiful trains, with a class system similar to, though less complex than, that of India. Their third-class was necessarily Spartan for conveyance of the poor Bantu and never used by Dutch or English. *Apartheid* was thus automatic. The third-class carriages were simply marked *Kleurlingen* or *Natives*, in Afrikaans or English respectively. They were what, in the southern United States, would have been called "Jim Crow cars", and so they are still. In America, old cars were used, not specially designed ones. Of course, where the work-grimed or the frankly unwashed travelled in large numbers, something of the sort was unavoidable. Even the plushy British railways had wooden-furnished coaches for miners' and quarrymen's "paddy trains". Pit-head baths were as yet rare. Car weights in, say, 1914 were very variable on standard-gauge alone: eighty tons or more for a massive American Pullman, half that for a German *D-Wagen* or a large English diner, and possibly no more than eight tons for some old fossil on a French or Spanish branch line. Running was equally variable. The Pennsylvania Railroad and the London and North Western Railway both claimed, with some justice, to furnish the smoothest travel in the world. One could also find the contrary on certain lines of both England and America.

We may notice here an interesting thing shown at the German Railway Technological Exhibition at Seddin in 1924; a sleeping car with single-berth compartments for first- and second-class passengers. The latter needed to be rather acrobatic, and by no means a claustrophobe, which may be why the arrangement did not go into large-scale production. But in the first-class compartment was the prototype of the American Pullman "roomette", which was to advance with singular success in the nineteen-thirties.

Observation cars were rare in Europe. In 1914, there was a solitary, very sumptuous, British-built Pullman observation car in Scotland. As Scottish weather, blowing westerly straight off the North Atlantic, is nearly as hazardous as the Scottish West Coast is beautiful, the rear end was entirely glazed. About the same time, some Canadian Pacific observation cars found their way to Austria, Tyrol, under very curious circumstances having to do with the Canadian Pacific steamships. (There was even an anti-British row about them in the German-language newspapers!) They ended their days in Italy long after World War I, again under very curious circumstances. One, perhaps the last, was at Voghera for breaking-up in 1961. Goodness knows who, if anyone, paid for them!

Apart from traditional car-classics, American, European and Eastern, there were many combinations of these. In Central America and in parts of South America one could see cars of absolute North American type, but with second-class seating very like Indian third-class of that time; back-to-the-wall each side and back-to-back in the middle. South American cars, often built in England or Scotland, and beautifully done-up in Argentine ox-hide, combined features of European and American practice, though the motive power might be, and probably was, as British as the cities of Glasgow, Newcastle or Manchester. South American day-cars looked British as to coachwork, but were seated American-style in most cases. We exemplify a contrary example, in a Scottish-built first-class carriage (*reservado*) on the Chilean Nitrate Railways, entirely European within but, outside, rather Old-American. Made as late as 1927, it perpetuated the clerestoried wooden body, which was already obsolescent in Europe.

There was a handsome smoking saloon which belonged to a twelve-wheel Pullman parlor car built in the United States for the London, Brighton and South Coast Railway in 1906. It was one of the last three American cars imported by an English company. The style was classic Pullman of the period, though the provision for smokers was much more liberal than in American smoking compartments of the time, which on Pullmans were combined with the men's lavatories. (Your non-conformist either washed in front of a row of cigar-impaled grins or else smoked amid half-bare dripping bodies!)

This compound duplex locomotive was used for heavy coal trains on the Baltimore and Ohio Railroad, 1903.

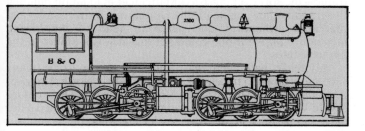

The *Matt. H. Shay's* wheel arrangement was known as triplex. It was used
for freight haulage on the Erie road.

The true rock-bottom of American car construction was that
chassis and body were made in one piece. The old English ideal
of body-on-underframe died hard.

FREIGHT ENGINES

Now to that important, and very different, thing which is the
haulage of freight! We have already noted the appearance of
the Mallet semi-articulated steam locomotive in Europe from
1889 onwards. America was cautious, but in 1904, J. E.
Muhlfeld, Chief of Motive Power on the great and venerable
Baltimore and Ohio Railroad, took it up for heavy coal
haulage, in an enormous double-0-6-0 (C-C) locomotive, of
which an example was fortunately preserved. It was an
immediate success. America took the Mallet to her ample
bosom, where it prospered like many immigrants. Europe was
ever cagey about it. The engine was sluggish, but immensely
powerful, and easy on the road. How odd that the very earliest
Mallet of which we have record was a tiny little double 0-4-0
tank engine which could do useful work on Decauville's
portable railways, useful alike on mobile industrial work and in
war! In America the Mallet became the tireless giant in heavy
traffic—the sort of thing J. A. Maffei had in mind on building
the grandmother-engine for the Gotthard in Switzerland—and
thereafter, whenever one heard or read of *the biggest
locomotive on earth* one could bet that it was a Mallet in the
United States!

Muhlfeld's had been that, and there were to be many
successors down to the end of steam traction in North America.
A prodigy was an engine called *Matt H. Shay* on the Erie before
World War I, which not only had two sets of four coupled axles
each but a third under the tender, making the combined

wheel-arrangement 2-8-8-8-2, with the type-name Triplex. There was nothing new in steam tenders; they had been made in both France (Eastern Railway) and England (Great Northern Railway) in the middle of the previous century to assist traction on starting and on heavy grades, but there was the same pitfall before them all. That meant too many cylinders for the boiler to supply, and all steam men have been ever agreed that, short of accident, there could be nothing worse than losing one's steam. Even the mighty "Shay" was not immune!

Another problem with these much-elongated locomotives was that of boiler-barrel length, and there were desperate attempts to make the boiler bend in the middle by what one can only describe as a bellows-connection, with the front half as a feedwater heater of sorts. One would think that this was asking for trouble, and unfortunately we have not met any man who had to do with such a thing—and would talk! It did not persist, but the Atchison, Topeka and Santa Fe Railroad certainly built this type, with six coupled axles in two sets, making it 2-6-6-2. Weight in working order was 308 tons, to which the twelve-wheel tender added 117 tons. Oil was the fuel.

Mallet locomotives appeared in many other parts of the world, though not on such a gigantic scale. British firms built them for South Africa, Burma and China. The type appeared on the Eastern Railway of France, and, from Maffei, on the Bavarian State Railways; all these on a much larger scale than in the early days of the Mallet in the Alpine countries. The same applied in Hungary. German works built broad-gauge Mallets for the Central Aragon and Zafra-Huelva Railways. In Scandinavia the Mallet was scarce, though an admirable narrow-gauge type by Atlas of Stockholm worked for many years on the .891 m. Dala-Norrsundet Railway in Sweden. Two survive in reserve.

THE BEYER-GARRATT

England produced a type of articulated locomotive which later was to be so widely and successfully used that Americans (who never bought it) sometimes called it the "British Mallet". This was the engine known down the years as the "Beyer-Garratt", for Herbert Garratt, an Englishman in Australia, invented it, and the great Manchester firm of Beyer, Peacock and Company acquired his patents.

"East Africa". Oil painting by Hamilton Ellis showing a Beyer-Garratt articulated steam locomotive on the Kenyan Plateau.

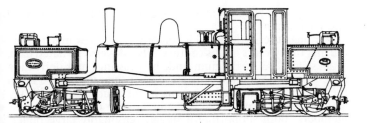

*Herbert Garratt's first articulated
locomotive, 1909.*

Basically, the Beyer-Garratt consisted of two distinct
locomotive units, each supporting through pivots and bearings
a proportionately large boiler on a girder frame between them.
Garratt's original idea had been to produce thus a powerful yet
flexible large-wheeled passenger engine, with no interference
by the height of the wheels in the dimensions of the boiler. He
had also ideas for a locomotive gun carriage, the guns being
mounted on top of the substantial tanks which were
immediately over the engine units and balanced the heavy
boiler between them. The express engine, with its wheels
arranged 2-4-0 + 0-4-2, was indeed built for the Sao Paulo
(Brazilian) Railway in 1915, but the very first of the type were
small double-0-4-0 locomotives for very narrow gauges; for an
isolated 2 ft. gauge of the Tasmanian Government Railways in
1909, and in 1911 for the vertiginous Darjeeling Himalayan
Railway in India. Few would have guessed that the
Beyer-Garratt of later years was to be a giant. Almost
invariable features of the type were outside cylinders with
radial valve gear of Walschaerts type. The little Tasmanian was
exceptional in having compound expansion, like the Mallets of
the period. She is preserved today by the Festiniog

*Kitson and Company of Leeds,
England, built this Meyer locomotive
for the Anglo-Chilean Nitrate Company.*

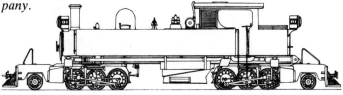

undertaking in North Wales, having gone half round the world and back again.

Various other articulated locomotive types saw much use in the first thirty years of this century. The Meyer type dated back at least to 1873 in Belgium and, by drawings and derivation (e.g. from the Semmering trials) earlier than that. Here again there were two motor bogies, as in the Fairlie, but with the leading one on a spherical pivot and the after one on transverse members, both supporting a single, reasonably large boiler. The main frames were those of the two motor units, which carried the couplers in consequence, while the boiler supported the tanks. There were separate exhausts, fore and aft. Such an engine was built for the Belgian Great Central Railway (that was the essay of 1873), but it was not until the 'nineties that the type really *arrived*, at the instance of Robert Stirling on the Anglo-Chilean Nitrate company's railway, who suggested it to Kitson and Company in England, who in turn built it.

In this version the boiler, tanks, bunker, in fact the "top works", were mounted on two parallel girders which rested on the bogies. These locomotives in Chile managed well on a seventeen mile line with a gradient of 1 in 25, about three-quarters of it additionally curved on a radius of 181 ft. The type achieved, in the present century, great success on other lines in the Andes, *viz*. both the Argentine and the Chilean Trans-Andine Railways, all on narrow gauge. The largest were in a set built, not for the narrow gauges of the Andes, but for the big broad gauge of Spain on the Aguilas-Lorca and Baza (Great Southern of Spain) Railway, a very considerable carrier of iron-ore. Kitsons held the Meyer patents, just as Beyer Peacock held the Garratt ones, and as the Great Southern of Spain was a British company until Spanish Communists got it, everybody was for a while happy. The engines were certainly a good investment. Other

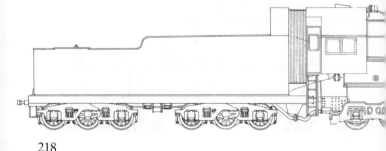

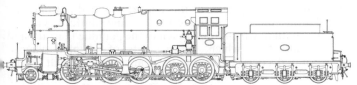

J. B. Flamme's 2-10-0 freight locomotive for the Belgian National Railway Company.

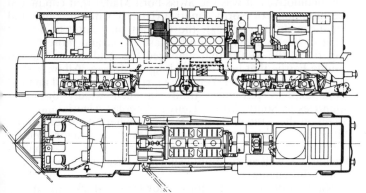

The Swedish 1650 HP Diesel Electric Bo-1-Bo Snowplow and locomotive type B.

In 1934, the Russian Railways built this stupendous 4-14-4 steam locomotive at the Voroshilovgrad works.

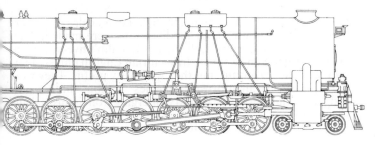

articulated steam engines of this period included not only the later Fairlies but the "modified Fairlie" and Maffei's "Garratt Union", both of which owed much to the genuine Garratt, and the Hagans type which, in two sets of coupled wheels, incorporated a derived drive to the rear set. It was built for the Prussian State Railways in 1893, and also, like the very different, pioneer Garratt, for the Tasmanian Government in Australia's Island State.

Summing up the distribution of articulated steam reciprocating locomotives, it was the Swiss-born Mallet that conquered the mountains of North America (latterly without compound expansion in most cases), the Garratt that came to possess African heavy haulage (South, East, West and more briefly Algerian), the Meyer that managed the Andes, and the old double Fairlie that held out for long in Mexico. All were slow and plodding engines except the Garratt, which could and did work express passenger trains in such widely dispersed countries as South Africa, Kenya, Angola, Algeria and even Spain on the Aragon Central Railway, where it succeeded the Mallet as it had done in South Africa. The Aragon Central company had a solitary Garratt express engine, a "double-Pacific", as well as the more usual heavy freight type. It was a wonderful engine.

Many railway authorities, however, stuck to heavy locomotives on a rigid, multi-axle wheelbase. The 0-10-0 engine was much favoured in Central Europe, by Prussia, Bavaria, Austria, Sweden, Italy, and perhaps most of all by Russia, where an admirable design, produced first before the first World War, was to be built in many thousands, notably as an all-work engine in the hard years after the Revolution, when large numbers were built by Nydqvist and Holm in Sweden and by many German Works such as Vulkan of Stettin and Humboldt of Köln-Deutz. These three firms all had the advantage, when it came to building broad-gauge locomotives, of access by waterway. From Lomonossov in Russia came staggering orders for 700 in Germany and Sweden. We show the Nydqvist and Holm engine, though the same drawings were used by all, and only minor modifications were made in later Russian construction under the first of J. Stalin's five-year plans in the late nineteen-twenties. In Victorian English phraseology, they were not beautiful, but they were very useful. They were also simple, straightforward engines without bothersome gadgets, suitable to a vast country emerging from industrial and social chaos.

In North America, the 0-10-0 locomotive was used only as a

(Above) *Mallet Type 2-10-10-2 built for the Atchison, Topeka & Santa Fe Railway, 1911.*

(Below) *4F Class 0-6-0 for the LMS, 1924.*
(Second from bottom) *The Javanese State Railways Hannoversche 2-12-2T, 1913.*

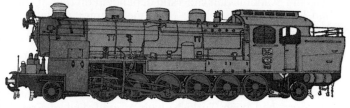

(Bottom) *French National Railways 241A Class 4-8-2, 1925.*

heavy switcher in the great freight yards, for pushing wagons over the humps of such tremendous outfits as Markham Yard, Illinois. The ordinary American freight hauler of the time was more likely to be 2-10-2, easier on the road than an entirely rigid engine. The 2-10-0 engine, of which a solitary de Glehn was built for Alsace-Lorraine Railways in 1904, was much liked in Europe, especially in Austria and the Balkans where it was to become even an express passenger type, at first on such lines as the Austrian Semmering, Tauern and Arlberg railways before their electrification, and latterly in Greece, where engines of obvious Austrian style, built by Skoda in Czechoslovakia after the break-up of the Empire, toiled and groaned through classic mountain places with the Athens portion of the Orient Express. One feels that they ought to have had names like *Leonidas*, *Alcibiades* and *Brasidas*, in a good Greek soldierly tradition, for they were most formidable and alarming-looking engines. By then, however, the naming of locomotives was mostly long out of fashion except in Great Britain, which ever loved this endearing custom.

Athens, be it added, was the last major European capital to be linked physically with the Continental network, apart from British and Scandinavian cities still dependent on train-ferries. It had long been possible to take one's choice of steam or electric trains between Athens and Piraeus, or to go loafing out on narrow gauge to Corinth and the coastal parts of the Peloponnese, but the south to north line through the great mountains, to link the Piraeus and Athens with Salonika via Larissa, was not completed until 1916, by the closing of the northern gap between Papapouli and Platy. Northern connections had been Turkish, with a good deal of French capital behind them, indeed foreign money went into most of Turkey's lines; French into the Oriental Railways, German into the Baghdadbahn, and British into the ponderously-named Ottoman Railway from Smyrna to Aidin. Their equipment reflected their origins.

This great 2-10-0 type of steam locomotive was, however, essentially a freight engine in Europe apart from these exceptions, and we show one of J. B. Flamme's for the Belgian State Railways' heavy coal traffic, after it had been rebuilt in the nineteen-twenties with improved front-end arrangements including double blastpipe and stack. Though more constricted by loading gauge than many European engines, these were in their day the largest and most powerful on the Continent. They were closely akin to Flamme's great Pacific type express engines of the same period, and like them

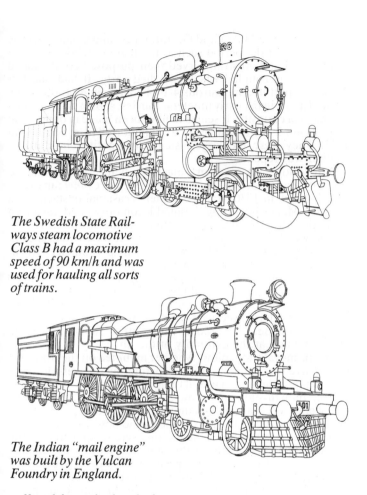

The Swedish State Railways steam locomotive Class B had a maximum speed of 90 km/h and was used for hauling all sorts of trains.

The Indian "mail engine" was built by the Vulcan Foundry in England.

suffered from the hand of war.

Austria, regarded popularly and not without reason as the European home of the big steam locomotive, essayed a 2-12-0 type in Gölsdorf's later years, and so did Württemberg which bequeathed it later to the German State Railway (a post-war corporation), but the coupling of six axles was fairly rare in Europe and uncommon even in North America. A quite recent example of the 2-12-0 was the A 1 of French National Railways (1948). The old Württemberg type ended its days on the Semmering line in Austria. Articulated grouping of

223

wheels, as in the Mallet and Garratt types, and in later electric locomotives of the period such as the famous "Crocodiles" of Switzerland and Austria, served their purpose better; indeed the very long coupled wheelbase died early. It had one last desperate puff in a monstrous 4-14-4 coal engine produced by Soviet Russia in the early nineteen-thirties, shown on p. 197 as a mechanical curiosity. It seems to have been made in opposition to a vast 4-8-2 + 2-8-4 Beyer-Garratt imported from England in 1933. They were the largest steam locomotives in Europe. Neither was repeated.

A few notes on accessory machines before we summarise general trends in steam design. Ever since the *Comet* ran over a wheelbarrow in 1830, trains had met occasional obstacles and hazards. Means for lifting derailed locomotives had advanced from primitive hand-crane to steam crane fairly early. As engine-weights grew, so side crane-lifts. It took two cranes to lift a locomotive bodily, with much preliminary work, especially if she were down on her side in the mud. In Europe, about 1910, ordinary lifting capacity might range from 20 to 60 tons; in the Americas it would be much more. Notable makers were Ransomes and Rapier in England and Bucyrus in the States. A modern American wrecking crane will have a lift of 250 tons at about 17 ft. radius with a lift of 50-60 tons on the auxiliary hook at the extremity of the jib.

The primeval snow-plow, whether mounted on the front of a locomotive or running as a separate heavy vehicle, was prow-shaped, and it is a useful thing to this day on routine snow patrol, though old style snow-bucking can be a hair-raising experience of charging headlong into the drifts with three locomotives behind the plow. The Leslie power-driven rotary plow, with a vast hooded, feathering screw, appeared in the United States very early in the 'nineties, and its use rapidly spread to most of the colder parts of the world. We show a modern, Swedish snowplow built by ASEA. Sweden was the first country to adapt electric power to the doughty rotary plow, on the Lapland iron ore line between the Gällivare-Kiruna area and Luleå on the Gulf of Bothnia and Narvik on the Norwegian North Atlantic coast. In the steam form, a high-speed geared engine was made to drive the rotor, supplied by an ordinary locomotive-type boiler inside the all-over cab, with a small tender in the rear. For Sweden, too, Henschel in Germany built a Diesel-engined rotary plow.

Now to ordinary locomotive practice of our rather elastic period: England, motherland of the steam locomotive, ever paid the price of pioneering in restricted clearances where her

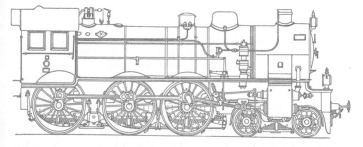

The Prussian State Railways Class S-10.

great main lines, most solidly constructed, thrust through the hills instead of skirting them. Restrictions plus magnificent formations caused multi-axle locomotives to be extremely rare. There had been a solitary ten-coupled tank engine on the Great Eastern in 1902, built for purely business reasons to show Parliament that a commuters' steam train could accelerate as rapidly as an electric rival then threatening the Great Eastern company. Another 0-10-0 locomotive, having four cylinders with simple expansion, in pairs with two valves and crossed ports, was built in 1919 by the Midland Railway for banking or pushing over the very steep Lickey Hills passage south-west of Birmingham. Otherwise the type did not appear on British rails until the advent of the 2-10-0 war engines in the nineteen-forties, some of which were to end their days on Dutch and even Swedish railways. (We have seen the famous *Longmoor* of these just twice; firstly when she went on the

The Highland Railways passenger engine Durn, *built in 1916.*

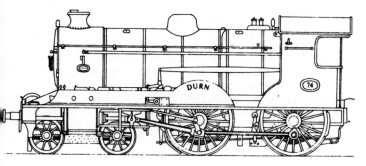

train-ferry at Dover, and the second time in the Netherlands Railways Museum at Utrecht.)

English conditions prevailed likewise in Scotland, whose railway system was geographically an extension of England though her old railway companies were distinctive, like the Austrian systems compared with the German though on a smaller scale.

Most of Great Britain's immense freight traffic was quite adequately shifted by six-coupled locomotives, with 0-8-0 or 2-8-0 for coal and other heavy minerals. Her equally astonishing passenger traffic (the London Brighton and South Coast, a fairly small railway, carried more passengers than the Canadian Pacific!) was dealt with over longer distances by modest 4-4-0 (the commonest), Atlantic, and 4-6-0 locomotives with axle-loads which some people would have found formidable, though the engines' aspect was modest by Continental, let alone American, standards. They were very compact, and usually most elegantly styled; further they were often painted (and kept very clean) in rich liveries of green, red or blue. As there were about 120 different British railway companies until 1923, there was plenty of variety at such great centres as York, Carlisle and Perth, which formed company frontiers. All the vastitude of America could not match those mechanical pageants of a British boyhood, though in Continental Europe there was a whiff of it at real frontier stations, and places like Antwerp and Frankfurt. One remembers with affection the fine brassy engines of the Low Countries, the majestic green ones of Bavaria, and the red-white-red collars about the necks of old Danes. As yet there were fewer national characteristics of style in electric locomotive design, save that they had pilots and central automatic couplings in North America and Mexico, and screw-coupling with side buffers on European standard and broad gauge.

The 4-6-0 steam locomotive was to be found from Western Portugal or Northern Scotland to the Urals and beyond; indeed it was one of the most widely-distributed passenger types over much of the world. There were such varieties as the Prussian P 8, one of the most numerous classes ever built, the famous four-cylinder simple engines of the Great Western in England, the Swedish State Class B, built over the decade 1909-19 (on the Stockholm-Västerås-Bergslagen in 1944!) and surviving to the last days of steam, the handsome Class A 2 of the Victorian Government Railways in Australia, and the slightly variable Indian "mail engine", British in origin, and built over a

Dr. Rudolf Diesel, 1858–1913.

(Below) *The* Lisboa *was one of 15 single-rail locomotives supplied by Sharp Stewart of Manchester, England to the Lisbon Tramway Company in 1872.*

period even longer than that of the Bavarian S 3/6 Pacific, a remarkable record.

Very many Europeans, and numerous visiting Americans, have seen the Prussian P 8, from Biscay to the Baltic and the Black Sea, for under the misfortunes of war it was vastly migratory. So it is a Prussian S 10 that we show here, a handsome four-cylinder simple express engine with the inside piston-valves worked through rockers by the outside valve gear. Four-cylinder compound 4-6-0 engines were common in France and in some parts of Germany, and in countries influenced by these. As hinted, four-cylinder simple 4-6-0 engines were most successfully built in England by George Jackson Churchward of the Great Western Railway, which was still turning them out in the war-battered nineteen-forties. Contrary to the Prussian S 10, these English engines had their outside piston valves worked off the inside gear. Also, they invariably had coned boiler barrels without domes, and Belpaire fireboxes, as well as that gloriously brazen Victorian aspect which distinguished the Great Western Railway for well over a century.

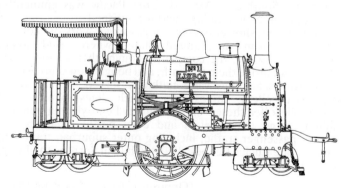

In both England and Scotland the 4-4-0 fast passenger engine was far from dying, however moribund it might be in the Americas or on most lines in Continental Europe apart from the Holland Railway guided by Ir. W. Hupkes, who was to become President of the Netherlands Railways through the heartbreak years to 1946. For an example let us take a design less well-known, for there were only two of them, designed by Christopher Cumming and built in a mighty hurry for the northern section of the Highland Railway, Scotland, in 1916. This hitherto remote line had suddenly found itself providing two railheads (Invergordon and Thurso) for the British Navy. *Snaigow* and *Durn*, as the sisters were named, worked innumerable naval trains. For all our qualification, we could scarcely call these "fast" on their final northern lap. British sailors in two wars called the journey *The Thirty-nine Stops*. Unusually, low platforms and splashers were combined with Walschaerts gear; aspect was rather austerely elegant, chaste as a bottle of malt whisky.

In North America the Pacific type locomotive had truly arrived for fast and ordinary passenger work. North of the 49th Parallel, the Canadian Pacific Railway cheerfully headed it into the Rockies, though the 2-8-2, originally built for Japan, helped more than somewhat. On the broad-gauge lines of South America, such as the Buenos Aires and Pacific in Argentina and the Sao Paulo in Brazil, both British-owned and using the Spanish and Irish gauges respectively, Pacific locomotives had shorter wheelbase, and narrow-grate Belpaire fireboxes instead of the wide-grate type of the true Pacific. In aspect they were English, yet with no counterparts in England; one of the curiosities of international capitalism blended with mechanical engineering tradition. On narrow-gauge lines, in South Africa, Japan and South-east Asia, the true Pacific was eminently suitable, as in New Zealand, and it flourished.

All the same, shorter engines for passenger work did not yet fade away from the United States where, said some un-American party, the companies bought their locomotives off the hook, drove them to death, and bought new ones. Many North American lines were faithfully served by mild little Moguls—excellent things for branch lines before the automobile killed these—and the mighty Pennsylvania went on building large ten-wheelers and even portly Atlantics for country passenger traffic, including outer suburban trains on what the late Lucius Beebe called its Child of Sorrow, the Long Island Railroad.

(Opposite) *Detail of* Clapham Junction *by Terence Cuneo.*

Lest there should be European sneers at the American custom of buying, driving hard, and chucking aside, let us recall that a ten-wheeler type of the Milwaukee Road, built by Baldwin in 1900, was hauling branch sections of the Hiawatha train about 1938, albeit much rebuilt and with a streamline casing on top. To be sure, there cannot have been much of the original engine left apart from the frames and the wheel-centres; certainly the original Vauclain arrangement of compounding on four cylinders, superimposed in pairs, had long vanished.

There was something of an English parallel. From 1946 we recall trans-Atlantic airline specials (two Pullmans and one baggage) being worked between London and the old Hurn Airport by Drummond eight-wheelers which had also been built in 1900, for the London and South Western Railway. Thanks to a few superheater elements, later installed, they behaved in very sprightly fashion!

From the lyrical to the practical; and harsh are the words of Apollo after the songs of Orpheus! The years immediately before war came in 1914 saw the first, stumbling entry of the Diesel locomotive. There were several essays with rail-motor cars running by compression-ignition on crude oil, in Germany, but in 1912 the Prussian State Railways had commissioned a large diesel locomotive to be built jointly by Borsig of Berlin and Sulzer Brothers of Winterthur. It appeared in the following year. Outwardly it suggested electric practice, with the wheel arrangement 2-B-2 or 4-4-4, with side-rod drives. Diagonal two-stroke engines gave direct drive through a jackshaft to the two coupled axles. There was no other transmission, and compressed air had to be used to bring the engines up to at least 60 r.p.m. before ignition began. The diesel was an ailing child, and war necessities brought an end to this experiment. Most mechanical revolutionaries were looking rather to increased electric traction.

Railroad traction by oil engines, thus initially dogged by transmission problems, belongs to later years, so let it briefly stand over. Rudolf Diesel was a cosmopolitan German, born in Paris in 1858, educated partly in Munich and partly in England. He published his *Theory and Construction of a Rational Heat Motor* in 1894. In 1913 he was invited to England for a talk on engines with the British Admiralty, but he mysteriously disappeared. His supposed body was taken from the Scheldt a long time after. Sinister stories were told, from different points of view, of *whose* secret agent was supposed to have tipped him over the ship's rail, and why. But it may have been that the

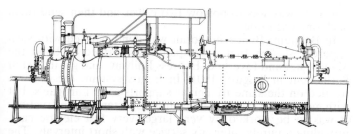

Lartigue's Monorail was used from 1888 to 1924 for a light railway in the south-west of Ireland.

unfortunate genius was leaning too far over when seasick, to take an involuntary header into the dark North Sea when nobody was looking. We are unlikely ever to know.

As a less sombre conclusion to this period, let us note monorails, which in the first decade of the century were attracting a good deal of attention. There was, of course, the Wuppertal line in Germany, the *Schwebebahn* with suspended electric cars, which is still with us and has lately acquired very modern rolling-stock, but the idea was much older than that. Even in the eighteen-sixties there was an arrangement for locomotives and cars to run on a single central rail with driving and balancing wheels to the roadway or ballast each side. This was more correctly a guide-rail line than a true monorail and it was quite workably applied to the Lisbon Tramways. We show one of the locomotives, built by Sharp Stewart, then of Manchester, in 1872.

Lartigue's monorail was incorporating an iron or steel trestle of triangular section with the running and traction rail on top and guide rails each side. It was demonstrated in Brussels, saw a very brief and unfortunate public service between Feurs and Panissières in France, and was used from 1888 to 1924 for a light railway from Listowel to Ballybunion in south-western Ireland. Never a goldmine, it was finished by damage in the Irish Civil War, repair of which was beyond the little company's means, but it was the longest-running of any steam monorails, and people travelled from distant places to see it and to ride in its rattle-drumming, pannier shaped cars. The locomotives had twin boilers, each side of the engine on the top rail. There were odd disabilities. It was impossible to convey a single cow; there had to be another to balance her on the other side of the rail. One way was to borrow two calves and then send them back one-each-side.

In Belgium, Behr's monorail was demonstrated with such apparent success that a company was formed to build the Manchester and Liverpool Express Railway, and got its Act of Parliament in August, 1901. In Behr's system, the rail was trestled on the same triangular supports as in Lartigue's, but with more guide rails and much more substantial construction. It was to be worked by very large, very fast electric cars, providing a purely inter-city service with short intervals. The London and North Western, and other companies, were much agitated, but the investing public fought shy and the line never was built.

Louis Brennan demonstrated a pure monorail whereon motor cars, balanced by large gyroscopes, ran with double-flanged wheels on one ordinary steel rail laid on the ground, or even on a steel cable. The railed version greatly excited people at an exhibition in London, in 1910. A similar car was sponsored by the newspaper king August Scherl and built in Germany at this time. Many said that the steam railway was imminently doomed. But it had yet a long and honorable way to go!

*The Canadian National Railways
4-6-4 Hudson.*

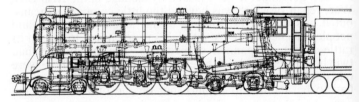

232

Chapter 7
Steam In Our Time

In the nineteen-twenties, the railroads, as an industry, and the railway train as a machine, first really began to encounter that dire Nemesis which is the breaking of established monopoly. We have seen this already in respect of the steam locomotive, but from the nineteen-twenties onwards it was the train itself. Hitherto it had been the only important form of mechanical land transport over appreciable distance. Now the motor and the aeroplane, rapidly developed under urgent war economy, moved in. The commercial motor was something very different from the rich man's private car, just as, after a second great war, the air-liner of our time is a very different thing from the old Fokkers, Junkers and Handley-Pages of the 'twenties and early 'thirties. That the train has survived to this last, though by no means everywhere, is a token of its strong superiority for certain sorts of traffic.

Certain other sorts of traffic have simply vanished, most rapidly in wealthy or war-winning countires where there is the additional urge, psychological as much as big-businesslike, to scrap the old and get on with the new. The last of Europe's great railway companies—those which maintained a monopoly in the British Isles until 1947, and certain important undertakings in Sweden—showed wisdom when they sold out to Government, even under onerous terms imposed by Socialist Administrations. One recalls angry scenes at the last annual general meeting of the London Midland and Scottish Railway, and has heard Swedish remarks about the rape of the Bergslagernas Jarnvägar which, to the last, was *a rather good business*. But the United States of America had an armour-plated *credo*, that of free enterprise. It has resulted in our time in the

The 4-6-4 Hudson was first built in 1927 for the New York Central Railroad.

persistence of company-owned railroads (or none) except in the new pioneer State of Alaska, and to achieve that business-happy object, many of the surviving railroad companies have been at great pains to shed their passenger traffic altogether. Not unreasonable, in some places! Even when we were in studenthood, an American fellow student (in Munich of all places) said then when he got home he was buying a *flivver*, which would solve all his own travel problems up to 400 miles, which to us meant the distance between London and Edinburgh. *Flivver* meant not a giant Cadillac; it meant Mr. Ford's T-Model. Germans had invented the motor-car. Frenchmen made it great. Americans, already, were making it part of the American Way of Life. Later results are just beginning to be a shade ironical.

There is our background to the sober mechanical history which must now ensue!

It is interesting that it was from Switzerland, home today of one of the most advanced railroad systems of the world, that we first heard, even in the nineteen-twenties, of what could only be called a sinister idea; that it would be a good thing to scrap the Swiss railway system (in the middle of Western Europe with all its international connections) and turn the old formations, yea, the great Gotthard itself, into motor roads. There were not many motors around then, but doubtless that could be remedied. The thing was that Switzerland's railroads were in a very bad way with an enormous deficit. The tourists had vanished in the war years, and of the surrounding countries, France and Italy were at war with Austria and Germany, so there was nothing to be had from neighbours' misfortunes, as there would be in a later war when Germany and Italy were aligned. Further, it had been very difficult to get foreign

The Canadian National Railways
4-8-2 Class U-1 was built in 1930.

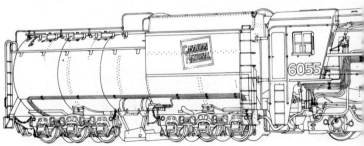

234

fuel. Switzerland's only fuel was falling water. The only Swiss railroad using it on a grand scale was the Bern-Lötschberg-Simplon, financed largely in France.

The Federal Government of Switzerland wisely cancelled its railways' unpayable debt, having embarked on a national policy of railroad electrification. It was begun, and ultimately done. A time was to come when in all south-western Europe (after the next war) they would be the only major railway undertakings to remain (in bankers' language) viable. The idea of transport being a national service, as against a commercial undertaking, was little in countenance as yet save in newly-communist Russia and grandly-capitalist Switzerland.

There was the background. In the foreground of mechanical history of that period are the interesting facts that Switzerland, full of water-power, went for that with willing capital expenditure, while Russia, with much unrealised wealth of oil, put into service some of the very first heavy Diesel locomotives on her long-neglected and war-battered railways. The Russian essay was not to bear much fruit for a long time to come, and ironically the most important railways in the Soviet Union were ultimately to go electric. But in the Swiss company went Italy, France, Germany, Sweden and Austria, as rapidly as they could afford from the middle nineteen-twenties onwards.

The rest of the railroad world carried on, in general, with coal-fired steam, though Russia used oil-fired steam in many places, as she had been doing to a limited extent for a long time already. Finland managed uncommonly well on wood fuel for many of her trains. Fiji and some other places happily ran narrow-gauge steam on sugar-cane trash, which was not nearly so absurd a fuel as some people might imagine.

In 1925, the London and North Eastern Railway, as

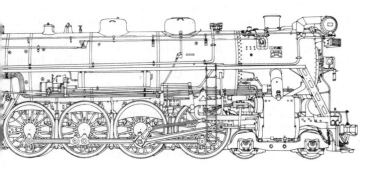

remarked earlier, celebrated a Railway Centenary by virtue of the Stockton and Darlington opening, and one of the events was a procession of locomotives and trains, ancient and contemporary. As we remember it, everything in that procession was steam-propelled except Stephenson's *Locomotion*, which had a petrol engine hidden in the tender and some old motor tyres acridly burning in the firebox for dramatic effect, and a petrol-electric railcar. A large North Eastern electric locomotive, built for a York-Newcastle scheme that to this day has never come off, was rather ignominiously towed past the stands by a little tank engine. It might be the afternoon of steam in those nineteen-twenties, but still it was Steam's Day!

THE HUDSON AND MOUNTAIN TYPES

At this stage, direct steam traction was very far from dead; indeed, it was kicking as well as alive. On some lines in the United States, the revered Pacific was being elongated into both the Hudson (4-6-4) and Mountain (4-8-2) types. The former had originated in two experimental locomotives by Gaston du Bousquet for the Northern Railway of France, back in 1910. They were intended for the Nord Express between France and Northern Europe. Their designer consequently named the type Baltic. It differed from the later American Hudson in having bogies at both ends, for the Hudson had a radial truck at the rear. This distinction seems not to have been noted in English-speaking countries.

As for the 4-8-2 engine, it had made several appearances in North America by the nineteen-twenties. Taking the Pennsylvania Railroad for once in a while on its own evaluation—*the Standard Railroad of the World*—we may remark that it had its first Mountain type in 1923 (Class M 1) built at Altoona under J. T. Wallis, and that following practical experiments with this, a modified design was worked out and 200 were ordered, 175 from Baldwins and the rest from Lima Locomotive Works, for heavy passenger and fast freight haulage, notably over the Allegheny Mountains between Altoona and Pittsburgh. True to Pennsylvania practice, a firebox of Belpaire type was used, but with a great combustion chamber extending over eight feet into the barrel (an old arrangement, justified by the extra value of firebox-heating surface, but rarely carried out on this scale). Again true to Pennsylvania tradition, the engines had very handsome lines.

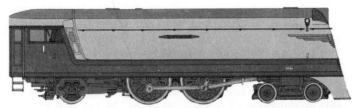

The 4-4-2 Hiawatha.

Gresley's A/4 Sir Ralph Wedgwood.

(Below) *New York Central Railroad's J3A Class 4-6-4, 1937, streamlined in 1941 to work the "Empire State Express".*

Union Pacific Railroad's 800 Class 4-8-4, 1937.

Dimensionally they were scarcely bantam-weights. The working pressure was 250 lb. per sq. in.; the cylinders (two) were 27 in. diameter with a 30 in. stroke, the distribution was by 12 in. piston valves, the boiler had 6,332 sq. ft. of combined heating surface, and the engine weighed 382,400 lb. without the tender, which came to 217,900 lb., in working order. In 1930, the company built twenty-five at Altoona, with fifty more by Baldwin and twenty-five by Lima (Class M 1a). We are not giving much of dimensions in this chronicle; those must serve as a sample of the big orthodox-type steam locomotive in North America in the nineteen-twenties.

Also in 1923, Canadian National Railways produced a very comparable Mountain type engine, and likewise based later standard designs on this. Canadian locomotive design continued to follow United States practice very closely, though not utterly, and the standard Canadian National "Mountains" appeared in 1929 for the Winnipeg-Edmonton run. In 1925, the type made its appearance on European main lines, on the Eastern Railway of France, which was followed by the Paris, Lyons and Mediterranean Railway. The type name was preserved. It was *Le Mountain* (not, observe, *Montagne*!).

Weimar-Republican Germany had, during the nineteen-twenties, the largest national railway mileage in the world after the United States and worked a great part of it with our old friends the P 8 and with eight- or ten-coupled freight engines, though for fast passenger traffic in the South there were the beautiful Pacifics of Munich and Esslingen, with larger versions being built in the North by such firms as Borsig of Berlin, and Henschel of Kassel. These also produced from 1924 2-8-2 locomotives for all sorts of trains, from the rather leisurely German *Schnellzug* of that time to fast freight, with heavy ordinary passenger service in between. Previously, the Saxon State Railways had built the first 2-8-2 express engines with really large coupled wheels, as in a Pacific. The English would have called the Reichsbahn type mixed-traffic engines, though this had nothing to do with that picturesquely awful thing the Spaniards called, in the way of trains, a *mixto*. (It generally meant a few old third-class carriages at the end of a caravan of local freight.)

For an example of a European 2-8-2 steam locomotive we give a remarkable engine which we first encountered at Liége in 1930. It was a design made for the old Luxemburg line, 121 miles long from Brussels to Arlon with a ruling gradient of 1 in 62, and was for express passenger service, to succeed the Flamme Pacifics. F. C. V. Legein, Flamme's successor, was

(Right) *The Kylchap blast-pipe.*

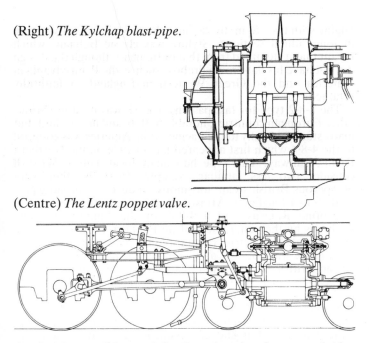

(Centre) *The Lentz poppet valve.*

the designer, and he used the American bar frames and Alco type piston valves while retaining many traditional Belgian features. The great firebox had four arch tubes (water-tubes supporting the brick arch) and the boiler carried a working pressure of 200 lb. per sq. in., which might have been made higher. Even so, the new Belgian Type 5 was the most powerful passenger type in Europe at the time. It was a simple

(Below) *A 4-6-2 two-cylinder express
locomotive, no. 3.1249, Chemin de
Fer du Nord, 1934.*

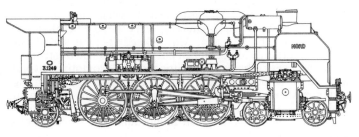

engine with two $28\frac{3}{4}$ in. by $28\frac{3}{8}$ in. cylinders (thus exactly square in longitudinal section). Styling was classic Belgian, which meant that it stemmed from that of Belpaire, though the design did not have his flat-crown firebox, such as the Pennsylvania in America and the Great Western in England so faithfully retained.

The 2-8-2 type for fast passenger work was used in various other European countries, notably in Italy and in Scotland, but many stuck faithfully to the Pacific, while America was going on to the 4-8-4 which first appeared late in 1926 on the Northern Pacific Railway. This type had many local names. We will consider two less familiar examples from the then very prosperous British steam locomotive industry; a broad-gauge (5 ft. 6 in.) Pacific by Armstrong-Whitworth for the Central Argentine Railway and a 4-8-4 for the Chinese National Railways by the venerable Vulcan Foundry (now part of the great English Electric organisation). This fine Argentine locomotive was notable, among many down the years, for her classic British lines in American proportions and on broad gauge (5 ft. 6 in.). She had three simple-expansion cylinders $19\frac{1}{2}$ in. by 26 in., $74\frac{1}{2}$ in. driving wheels and a working pressure of 225 lb. per sq. in. The Vulcan Foundry engine for China was not, for all her myriapod aspect, an absolute giant of the American sort, but she exemplified a fine combination of adequate power with remarkably low axle-weights. Her contours, to be sure, were tremendous. In China, as in Russia and the Americas, there was plenty of room to build locomotives upwards and outwards. In England, the problem was that of getting them to the docks for shipment. They could not go by rail in such sizes as these. Terrified motorists sometimes encountered them, at strange hours, on vast, crawling motor carriers. Armstrong-Whitworth was fortunately close to the river, but Vulcan Foundry was not.

For in America too, steam was fighting a doughty action against the upstart oil engine, even though, as some party said at the time, the American rail-roads were selling out to General Motors. *Certes*, G.M. established what it called its Electro-Motive Division, making very large Diesel-electric locomotives! The Chicago, Milwaukee, St. Paul and Pacific Railroad certainly aimed at fast passenger haulage by steam; at first with a very advanced Atlantic type of locomotive covered with an air-smoothed rather than a truly streamlined casing, and then with the magnificent Hudsons (Class F 7, streamlined, built in 1938). The train was called *Hiawatha*, after that noble, mythical, honorable red man immortalized in English verse

Built in 1935, by Borsig, for the German State Railways, this streamlined 4-6-4 steam locomotive made a recorded run of 200.4 km/h in 1936.

by that most nice Yankee Henry Wadsworth Longfellow. *Hiawatha*'s cars also looked very nice in the bright yellow and red which the Milwaukee company laudably perpetuated for many years in a generally drab world.

In the steam locomotive, as in other engines then, since and henceforward, quite a lot of the streamlining was internal, as in steam and exhaust passages and, for that matter, in our own internal organs. The high spots of speed with entirely orthodox, streamlined, steam locomotives were achieved by two engines, English and German respectively. One of Gresley's, the *Mallard*, which had been modified as to the front end with the French Kylchap exhaust arrangements, having a double chimney, attained 125 miles an hour during brake tests on July 3, 1938, with a very brief burst at 126 on the dynamometer-car's chart. In May, 1936, one of three experimental Hudson-type German locomotives (Reichsbahn Series 05) had shown 124.5 on the dynamometer chart. These fireworks were the high rockets of the steam reciprocating locomotive. The British Mallard and the German 05.001 are preserved at Clapham

Northern Railways of France rebuilt this as the 3.11 Class 4-6-2 in 1934 (SNCF Class 231E).

241

and Nuremberg respectively for those who knew and loved the prime of steam and for those who will have never known it. The German "O-Fives" lasted inservice until 1958; the British engines, forming a larger class, worked into the middle-nineteen-sixties on Scottish services.

The Kylchap blast-pipe, noted previously, combined features of the Chapelon arrangement in France and of Finland's Kylälä blast-pipe, hence its name. It incorporated four nozzles or jets, usually paired below a double chimney, making eight nozzles in all. It was one of the many improvements in exhaust and front-end arrangements that were being made about this time, perhaps most notably by André Chapelon. There were indeed many experiments at this time in steam locomotive development. The noble old engine was fighting for its life against the electric motor and the internal-combustion engine. One offensive, not a very happy one, was with the use of very high pressures, as with the Schwartzkopff-Löffler and Schmidt types from Germany, certain quite academic designs by the Delaware and Hudson Company in the United States, and a remarkable 4-6-4 express engine in England wherein Gresley of the London and North Eastern made an immensely expensive adaptation of the Yarrow marine watertube boiler. With the Schmidt type, which included a closed-circuit system of tubes—steam to heat steam—there were some blow-ups with tragic results. It was tried in Germany, England, France and Canada.

Piston valves having largely superseded slide valves, there was a vogue for poppet valves, following motor practice and sometimes worked by rotary-cam instead of radial valve gear. Lentz (in Austria) and Caprotti (Italian) poppet valves were about the most successful and were widely used, though the piston valve scarcely retreated.

An interesting arrangement of the early 'thirties was the Cossart valve gear, now shown in a rebuilt Northern of France Pacific. The engine had been one of Bréville's improved de Glehn-du Bousquet type of 1924 and was thus modified by his successor Lancrenon, as a four-cylinder simple engine. Outside arrangement of the gear is obvious from the drawing, but there were four valves for each cylinder, in pairs at each end, and of the piston-type set vertically, rising and falling through levers and springs, with a rotary camshaft. This particular transformation seems to have been unhappy, but later use of the Cossart system of distribution was justified by many years' service.

Otherwise, this Bréville-Lancrenon locomotive is notewor-

A mechanical stoker was used by the Polish State Railways.

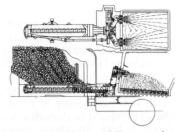

(Below) *The turbine locomotive built under the Zoelly patents.*

thy, like early engines on the Northern Railway of France, for the use of a long, narrow firebox, such as the Great Western Railway invariably favoured in England. Both companies achieved very fine performances with what Americans considered small engines. The Great Western boasted awhile, in the early nineteen-thirties, *the fastest train in the world*. The claim was quite authentic for regular service: Swindon to to London (Paddington), 77.3 miles in 65 minutes. On June 6, 1932, it was done in 56 min. 47 sec. (engine *Tregenna Castle*). This was a publicity boost. Rather slower services with much heavier trains between London and Plymouth were really much more interesting, with a mountainous road in the West Country. The Great Western engines, like the French ones, were among the best for their modest size in the world.

Many appliances were going to improve the basic steam locomotive still. The principle of thermic syphons in firebox was ancient (cf. J. H. Beattie in the eighteen-sixties). The modern Nicholson form was applied on certain American locomotives and in England (Gresley and Bulleid on the London and North Eastern Railway) at the beginning of the nineteen-forties. Absolutely essential, of course, was the big boiler supplying plenty of steam on a high pressure. Past were the days when partly through the need for keeping down axle loads, what had otherwise been very good engines were spoilt by boiler inadequacy, as in the eighteen-nineties.

In the late nineteen-thirties, such firms as the American Locomotive Company, and, as noted, General Motors Electro-Motive Division, were building larger and more powerful Diesel electric locomotives for both fast passenger

and heavy freight service in the United States. Certain American railroads stuck faithfully to steam for a very long time. The New York Central's legendary Twentieth Century Limited continued to roar along the water-level route behind its magnificent Hudson. The Norfolk and Western Railway, a very heavy coal carrier, showed quite fanatical loyalty to steam, even scrapping electric traction which it had installed before 1920. It is interesting to recall that as late as 1957, when the Diesel conquest of America was far advanced, the Norfolk and Western had 408 steam locomotives (including one terrific turbine-electric unit) called *Jawn Henry* and but sixty-two Diesel-electric. At that time there was no more steam on the New Haven, the Missouri Pacific, and many other Class I railroads. The New York Central had 122 steam and ninety-three electric locomotives out of a total of 2,168. Making European comparisons, in 1955 British Railways, under State ownership for the past decade, had still 17,955 standard-gauge steam locomotives against 532 of all other sorts, including 407 Diesel-electric and but seventy-one straight electric. The disappearance of classic British steam, little over a decade later, seemed unthinkable! West Germany, in 1956, had 9,533 steam, 525 electric, and 223 Diesels of all sorts. America undoubtedly was the first great country to be possessed by the Diesel.

Her steam locomotives had been vast things over many years. In North America, and in other countries of mighty steam engines, it had long got beyond the power of any long-suffering fireman to shift coal into those cavernous fireboxes, and the steam-driven mechanical stoker came to his aid, or rather to *their* aid where two firemen were needed to maintain steam. Its principle was that of one of the senior mechanical appliances, the *Archimedean screw* (third century B.C.). As so many American examples have been illustrated in past books, we show a diagram from the Polish State Railways. But yet it was in America that the appliance became universal on all big engines, and widespread in parts of Africa and the East. British

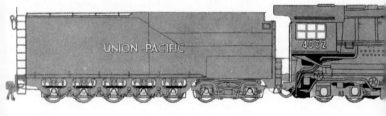

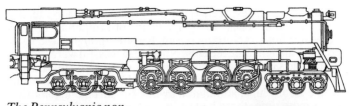

*The Pennsylvania non-
condensing turbine
locomotive.*

(Centre) *Soviet Railways'* Joseph
Stalin *Class 2-8-4, 1932.*

home railways never took it up. Their engines were necessarily small from first to last, and even some Beyer-Garratts on the Midlands-London coal trains went only as far as self-trimming bunkers which brought the fuel within easy reach of the strong man with the shovel.

Pulverised fuel, as an alternative to low-grade stuff, was tried at the expense of elaborate machinery for firing. Old Archimedes again was an essential contributory inventor. One should remark great progress in the use of pulverised fuel on the East German State Railway after World War II.

*Union Pacific's 4000 Class
4-8-8-4 "Big Boy".*

Boiler accessories, such as water purifiers (called top-feed in England) went back to Wagner in pre-Imperial Prussia. Feedwater heaters were increasingly used, though these again had quite ancient antecedents. Notably successful were those of Knorr (the German air-brake firm) and A.C.F.I. in France.

Experience of even torque, as with electric motors and turbine steamships, stimulated awhile the turbine locomotive. At a remote time, there had been a plucky but fruitless attempt in the United States to make a steam locomotive go by an enclosed reaction-turbine driving through belts. Ramsay's turbine locomotive experiments in Scotland and then in England bear mention. But there was brave work in Central and Northern Europe in the early nineteen-twenties. The Swiss Engine Works of Winterthur, and Krupps of Essen, both built notable turbine locomotives under Zoelly patents, using condensers as in marine practice. Rather oddly, it was the non-condensing turbine locomotive, with the usual open exhaust, that was to be a practical proposition in public service, while the huge, tender-mounted condensers were found useful for ordinary reciprocating locomotives in South Africa, and by both the Russians and the Germans, especially when they were both fighting and moving traffic in waterless country.

The Pennsylvania Railroad's straight turbine locomotive of the nineteen-forties followed on a grand scale those which had successfully worked in Sweden and England in the 'thirties, and were still going well. She was arranged 6-8-6. A Pennsylvania reciprocating locomotive of 1940 had been 6-4-4-6. As in the European engines, there was a main turbine for forward running (maximum speed 100 m.p.h.) and one for reverse running (25 m.p.h.) on the right and left sides respectively. Drive was through double reduction gears and quills to the two inner coupled axles, with roller bearings throughout, including the side rods. Pressure was 310 lb. per sq. in. (285 lb. at turbine inlet).

There is little doubt that had not the oil industry come to dominate world commerce and politics, far-advanced methods of using steam on the locomotive itself would have become commonplace wherever it was uneconomic to use it in central power-stations for electric traction and where water-power was weak. The Germans in particular made most interesting practical experiments with what we might call the steam-motor locomotive—i.e. one with geared high-speed engines.

As it was, the steam locomotive fought her last actions in what might have been called conventional armour, that of the reciprocating engine with direct drive and with such boiler and

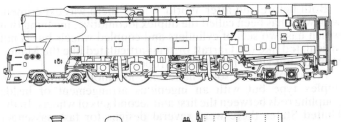

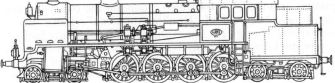

(Above) *The Pennsylvania Railroad's
4-4-4-4 Class TI.*

(Centre) *Netherlands Railways 4-8-4
tank locomotive.*

front-end improvements as mechanical science had already
produced. On very heavy haulage the giant American Mallets,
ever more gigantic, pursued their dogged way, and, in what we
still tend to call the Old World, especially in the African
countries, there was the Garratt. The classic American Mallet,
though she was not without her rivals even in size, was the great
design carried out by Alco for the historic Union Pacific
Railroad in and after 1941. A locomotive of orthodox type,
having a working pressure of 300 lb./sq. in., an evaporative
heating surface of 5,755 sq. ft. plus a super-heating surface of
2,043 sq. ft., a grate area of 150.3 sq. ft.; a locomotive with four
cylinders $23\frac{3}{4}$ in. by 32 in. (12 in. piston valves with 7 in.
maximum travel) driving sixteen 5 ft. 8 in. coupled wheels with
a four-wheel truck fore-and-aft; a locomotive weighing 772,000
lb. excluding her 348,000 lb. tender (two-thirds loaded) on five
fixed axles and a leading two axle truck, was not the sort of
steam engine one found moving up and down most people's
Mountain Divisions, even in North America. There were two
lots; road-numbers 4000-4019 and 4020-4024. The basic figures
above apply to the last five. These were the celebrated *Big
Boys*.

Both before and after World War II there was experiment with what has been called the duplex locomotive, meaning that which has two sets of cylinders and coupled wheels but is not articulated like the Garratt or semi-articulated like the Mallet. It made but slight impact in Europe.

France produced some handsome locomotives of apparent duplex type but with an ingenious arrangement of inside coupling rods between the first and second sets of wheels. In the United States there were several designs for fast passenger engines arranged 4-4-4-4 which, adapting the old type-name, might have been called "Double-American". There was the remarkable *George H. Emerson*, which incidentally had very handsome lines. The Pennsylvania Railroad's Class T 1, which first appeared in 1942 and ultimately numbered fifty-two engines, sang one of the more majestic swan-songs of railway steam. (Statement made with due respect to the Norfolk and Western, and several other undertakings!) The engines were designed for a normal maximum speed of 100 miles an hour at 20 per cent cut-off and 295 lb. boiler pressure. Maximum horsepower was 6,552 (indicated) and 6,110 (drawbar). Tenders were by now enormous. That of Class T 1 ran on two eight-wheel trucks, with the waterscoop between. Engine weight was 502,200 lb.; and of tender, 442,500 lb.

North America was the land of the giant passenger and freight engine; the steam tank locomotive was relatively little known there (or, for that matter, in the Soviet Union). There *were* indeed some large steam tank engines, 4-6-4 or 4-6-6, on certain commuting passenger services which had not been electrified, as on the Boston and Albany Railroad, and about Montreal on both Canadian National and Canadian Pacific services.

French National Railways 4-6-4 Class
U locomotive, 1949, used in the
Northern Region.

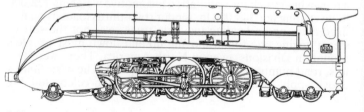

(Above) *South African Railways' 23 Class 4-8-2, 1938, built in Germany.*

(Centre) *Spanish National Railways' 3100 Class 2-10-2, 1943.*

(Below) *US Army Standard 0-6-0T, 1942, was built for war service in Europe.*

In many parts of Western Europe, however, there were tank engines galore; ancient, modern, diminutive, or even big to American eyes. Maffei in Munich built some tremendous Mallet tank engines as helpers on heavy grades in Thuringia. In the nineteen-twenties certain of the old British companies had built very fast-running and indeed elegant 4-6-4 tank engines for non-stop business passenger runs of fifty miles or so between London, Glasgow and Manchester, and their neighbouring coast towns. Then much larger 4-8-4 tank engines were built for Spain, France and most notably for the Netherlands Railways which used them for heavy coal haulage in and about Limburg.

One of these we show now. They were the largest and most powerful of their type in the world. A product of 1930, they were simple, reliable and very competent. One is now preserved at the Netherlands Railways Museum, Utrecht. There were four cylinders, all in line with drive to the leading coupled axle. Noteworthy was the combination of bar frames with an almost English outline. The stack had a bright copper cap, a Victorian feature favoured by the Netherlands, as well as the Great Western Railway in England, and endearing to the eye of an artist, as to an old-time locomotive man.

While in North America, the steam locomotive retreated in the way of mechanical evolution spurred by oil-business interests, in much of Europe, during the nineteen-forties, there was a massacre of another sort. In their thousands, steam engines fell to the waste of war, whether by aerial bombing or by sabotage. To take but one example, seventy per cent of effective steam locomotives in Austria were, officially at any rate, either severely damaged or quite destroyed, and (mark this!) a steam engine was the most resilient and least fragile of all locomotives. She could limp home after mischief that would have crippled anything else. One should add that no country could beat Austria at the useful business of having locomotives in reserve, sometimes with long grass about their wheels, but sound when needed!

It was in Austria that one of the last great contributions was made to steam locomotive engineering. Giesl-Gieslingen's improvements in front-end design and, particularly, in exhaust arrangements, were what earlier had been called a "break-through". But apart from the improvement of many existing engines and the better equipment of a few new ones, they came too late.

Many splendid locomotives were built in those fateful 'fifties which saw the twilight of European steam. In Italy, land of so much electric traction, the Franco-Crosti boiler revived in modern form something like Petiet's arrangement from long ago, even to rearward smokestacks, and tried early on a veteran "back-to-front" 4-6-0 of Plancher's type. Germany and England both gave it fair trial. In Ireland that unconventional-conservative New Zealander Oliver Bulleid—once of the English Great Northern and later of the English Southern—made a brave attempt to produce an advanced steam locomotive running on turf. He, like ingenious Germans before him, had great ideas about steam-motor locomotives, multiple-cylindered and incorporating sleeve-valves as well as much else that old shellback engineers considered unconstitutional and thus anathema.

Let our last bouquet go to France, country of Marc Séguin, Jules Petiet, Baudry, du Bousquet, and André Chapelon; adoptive country of Tom Crampton and Alfred de Glehn; nation of fascinating steam! We show now French National Railways Class U, one of the last of the great French compound express engines, built in 1949 by Corpet, Louvet and Company. The chief designer was de Caso, heir to the work of such great Frenchmen as Chapelon on the Orleans and Bréville on the Northern. In her design there were the best in tradition and also the best in novelties. Drive was divided, with high-pressure cylinders in a single casting driving the first coupled axle, and the two low-pressure cylinders outside, driving the middle one. Piston valves and Walschaerts' gear were revived. S.K.F. roller bearings were used, and lubrication was by automatic pumps. There was a mechanical stoker. This was not a big engine by North American standards, but with 300 lb. per sq. in. boiler pressure, making 270 lb. in the high-pressure steamchest and 75 lb. in the low-pressure, the initial engine on her trials between Paris and Lille easily equalled the speed performance of the Diesel *autorails* while taking loads up to 500 metric tons at 87 m.p.h. maximum speed. High average speeds were much regarded in France, without high maxima.

Specialised steam locomotives remained few in type. The rack-and-pinion mountain engine, pioneered by Marsh in America and Riggenbach in Europe, is still with us, in a few places like Mount Washington in New Hampshire and the Rothorn in the Alps, though rackrail Diesels now growl up to the tops of Pike's Peak and Monte Generoso.

For work over rough roads, especially lumber lines in North America, the Shay locomotive was widely used over many years. It was probably the ugliest Caliban of steam, with its boiler off the centre-line to balance a vertical engine driving a universally jointed propeller-shaft to two or three motor-trucks, but it was an ingenious machine, and extraordinarily useful at that within its limited sphere. Rivals were the Climax and Heisler geared locomotives, but the Shay remained the old favourite. Fireless steam locomotives, with heavily-lagged steam reservoirs charged at strict intervals from a central supply, originated for steam street-car lines in New Orleans, back in the 'seventies, and many years later they were found useful about mills and oil-refineries where open exhaust might cause disaster on a grand scale. Germany in particular produced many excellent examples in the present century.

With these humble housemaids of steam locomotion, we must leave what long bygone Englishmen called the *Steam Travelling Engine*. In its day, lasting about a century-and-a-half, it has been not only one of the most important of transient machines, but one truly best-loved, both by those who had to do with its honest reliability and by those to whom it simply appealed to the senses, as the great sailing ship, the windmill, the beautiful bridge, and a few other purely functional things have appealed.

We have called this *The Lore of Steam*. In lore, there is no place for prophesies, unless they are old ones on record, like George Stephenson's on electric power. We can only watch tendencies, which ultimately are governed by technical advance but are also swayed by business fluctuations, whether capitalistic or collective, by politics of all sorts, and most absurdly by fashion. There are ever old loves and old hates, for everything that is loved by some is hated by others. The old British ruling class in the first half of the nineteenth century—especially the English—hated the rail because it invaded the privileges of property and imposed a sort of policing on those who travelled. The old British proletarians had their horizons widened, but found themselves being brusquely pushed around. The British bourgeoisie of the same time were most content, and many of them became much richer than they had been. All classes in America (which was nominally classless) welcomed the train as the bringer of progress, yet were the first to turn against it in their sturdy individualism when the motor car gave them mechanical independence in travel. Prussia found in the rail a strategic asset, and used it with energy to this end. Having sojourned in Germany in the nineteen-twenties, we never encountered that latent hostility to the railway industry which in England had never died out.

In our time, the local country railway has withered away in

many places. The train has vanished from such small and isolated countries as Cyprus, Mauritius, Barbados and Jersey; all islands. The position in Venezuela is odd; the once spectacular La Guaira and Caracas Railway has been gone some time; higher up the country, great ore trains are running where no trains were seen a few years ago. More great ore trains are now rolling down Labrador where no trains ran before.

For the carriage of solid minerals by land, the train is as yet the only thing, just as for the movement of city millions it is the only thing. For close inter-city traffic it remains the best thing. For airport access it is a very good thing, whose virtues might have been appreciated earlier. In such traffic, the train will prevail, and serve, through foreseeable time. Just what form that train may take is a subject for lively discussion. Monorails, as yet, are footling things. The best is still the ancient one in Wuppertal. Completely automatic railways have been with us, though unobtrusively, for some time. The application of electronics advances. Traction by linear induction is under study and practical experiment though British experiments so far have shown it to be uneconomic at speeds under 200 miles (or 320 km.) per hour. These things are not yet Lore.

East German Railways' 01.5 Class 4-6-2, 1962, is a reconstruction of the pre-war Pacific Class.

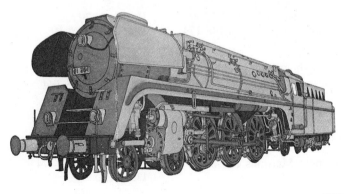

Index

Numbers in italics refer to illustrations.

255